Optimal Design of
Flexible Manufacturing Systems

OPTIMAL DESIGN

OF FLEXIBLE MANUFACTURING SYSTEMS

Dem Fachbereich Rechts- und Wirtschaftswissenschaften der Technischen
Hochschule Darmstadt zur Erlangung der Würde eines Doktors der
Wirtschaftswissenschaften vorgelegte Dissertation

von Dipl.-Wirtsch.-Ing. Ulrich A.W. Tetzlaff aus Frankfurt am Main

Referent	:	Prof. Dr. Horst Tempelmeier
Koreferent	:	Prof. Dr. Wolfgang Domschke
Tag der Einreichung	:	19. Januar 1990
Tag der mündl. Prüfung	:	3. Mai 1990

D 17

ISBN 3-7908-0516-5

Physica Verlag, Heidelberg 1990

Ulrich A. W. Tetzlaff

Optimal Design of Flexible Manufacturing Systems

With 30 Figures

Springer-Verlag Berlin Heidelberg GmbH

Series Editor
Werner A. Müller

Author
Dr. Ulrich A. W. Tetzlaff
Walter A. Haas School of Business
University of California at Berkeley
350 Barrows Hall
Berkeley, CA 94720 / USA

CIP-Titelaufnahme der Deutschen Bibliothek
Tetzlaff, Ulrich A. W.:
Optimal design of flexible manufacturing systems / Ulrich A.
W. Tetzlaff. - Heidelberg: Physica-Verl., 1990
(Contributions to management science)

ISBN 978-3-7908-0516-1   ISBN 978-3-642-50317-7 (eBook)
DOI 10.1007/978-3-642-50317-7

© Springer-Verlag Berlin Heidelberg 1990

Originally published by Physica-Verlag Heidelberg in 1990.

7120/7130-543210

Meinen Eltern - To my parents

ACKNOWLEDGEMENTS

I would like to thank all the people who have made this work possible.

First and foremost, my advisor Prof. Dr. Horst Tempelmeier, whose help, direction, inspiration and timely comments were invaluable.

Furthermore I would like to thank Dr. Armin Gohritz, member of the board of Dörries Scharmann GmbH, for giving me the chance to carry out this research project during my time with the company. His supportive confidence in my work gave me continued encouragement.

I want to express my gratitude also to Prof. Dr. Wolfgang Domschke for his advice and his willingness to be a member of the reading committee.

Also special thanks to my friends Klaus-Peter Kegel, Heinrich Kuhn and Dr. Erwin Pesch for their critical comments and advice, as well as to Prof. Dr. Andreas Drexl for his supportive suggestions during the early stages of this research.

Thanks must also be expressed to Detlef Mierswa, systems engineer at Dörries Scharmann GmbH, for many useful discussions.

Additionally I would like to express my special thanks to my colleagues Thomas Endesfelder, Wilfried Hauck, Reiner Hoenig, Birgit Schildt, Hans-Jochen Schmitt, Dr. Stefan Voß and Volker Weber, who helped me with many problems during my work.

Finally many thanks to the secretaries Ute Jahn and Ruth Jahn and the students Harald Deprosse, Christine Meier, Andreas Mengen, Volker Müllers, Alexander Pollklesener, Hans-Ulrich Schüler and Oliver Wild for their patience and support.

CONTENTS

PART A: INTRODUCTION

1. Overview

Since the late seventies flexible manufacturing systems have become increasingly popular in manufacturing. The possibility of manufacturing a variety of different part types on a versatile automated system seems to be especially attractive for an environment with low to medium-sized yearly production requirements.

However, the capital investment for such a system is considerable, which raises two major questions:

- How can the investment be economically justified?

- How should the system be designed so that production needs are satisfied in an optimal way?

Until now the first question has attracted most of the researchers' attention. There are a large number of articles about different justification models[1], in which a continuing controversy about the applicability of traditional financial evaluation methodologies can be observed.[2] The second question, however, has been neglected or only treated in a rudimentary way. This seems surprising, considering that a sound decision on investment in a flexible manufacturing system can only be made if it is possible to design an optimal system which exploits all the advantages of flexible manufacturing. The author believes, that the discussion about the justification of flexible manufacturing systems is redundant as long as the question of how to design an optimal system is not answered satisfactorily, i.e. in order to answer the first question a solution to the second must be found.

Therefore this thesis is devoted to the second question and tackles it as follows:

- the design problems are analyzed in detail,

- a planning concept is given which structures these problems,

- possible tools for the design process are described,

- existing optimization models for the design are presented within the framework of a classification scheme,

- and optimization models based on mathematical programming, newly developed by the author are presented. Here the major contribution consists in providing two models with their solution procedures for the equipment selection problem considering dynamic aspects of system behavior, i.e. queueing processes and part flow interactions in the system. This problem of equipment selection can be described as follows: Given

---

[1] to mention only a few: Burstein, M.C., Talbi, M.: Economic Evaluation for the optimal Introduction of Flexible Manufacturing Technology under Rivalry, in: Annals of OR, 3(1985), pp.81-112; Hutchinson, G.K., Holland, J.R.: The Economic Value of Flexible Automation, in: J. Manuf. Syst., 1(1982)2, pp.215-218; Horváth, P., Kleiner, F., Mayer, R.: Dynamische Investitionsrechnung für flexible automatisierte Werkzeugmaschinen, in: DBW, 47(1987)1, pp.69-84; Kulatilaka, N.: Capital Budgeting and Optimal Timing of Investments in Flexible Manufacturing Systems, in: Annals of OR, 3(1985), pp.35-57; Meredith, J.R., Suresh, N.C.: Justification techniques for advanced manufacturing technologies, in: IJPR, 24(1986)5, pp.1043-1057; Nelson, C.A.: A scoring model for flexible manufacturing systems project selection, in: EJOR, 24(1986), pp.346-359; Primrose, P.L.: Evaluating the 'intangible' benefits of flexible manufacturing systems by use of discounted cash flow algorithms within a comprehensive computer program, in: Proc. Instn. Mech. Engrs. Vol. 199 No.B1 (1985), pp.23-28; Suresh, N.C., Meredith, J.R.: Justifying Multimachine Systems: An Integrated Strategic Approach, in: J. Manuf. Syst., 4(1985)2, pp.117-134

[2] see for example: Michael, G.J., Millen, R.A.: Economic Justification of Modern Computer-Based Factory Automation Equipment: A Status Report, in: Annals of OR, 3(1985), pp.25-34, Miltenburg, G.J., Krinsky, I.: Evaluating Flexible Manufacturing Systems, in: IIE Trans., 19(1987)2, pp.222-233

is a part family and its necessary production volume to be produced on a flexible manufacturing system. To allow the selection of equipment, alternative routes through the system must be provided, i.e. alternative sets of workloads on different machine types, on which each part type (or part group) of the part family can be produced. Now the optimal configuration (number of machines of each type, transportation vehicles, pallets etc.) must be found, so that equipment costs, in-process inventory costs (and if necessary, operating costs) are minimized, while given production requirements for the part family are fulfilled. So far the existing models in literature, which consider dynamic system behavior allow only optimization of the capacity of a system for given types of equipment, but do not include the selection of equipment, i.e. the inclusion or exclusion of certain types of equipment. However, in practice this aspect is of major importance, because the flexibility of modern CNC-machines gives the system designer the opportunity of choosing between different types of equipment with different costs but very similar capabilities.

This dissertation is divided into three parts and consists of nine chapters. In the first part an introduction to the subject is given. It starts with this overview and continues in the second chapter with a few introductory definitions. A flexible manufacturing system and its characteristics are described. Additionally, some problems relating to flexibility are discussed. Chapter three presents an overview of planning problems. They are divided into design and operational problems. The latter affect an already existing system and therefore are only addressed briefly. However, the main area of interest here are design problems, i.e. the problems involved in the design of a flexible manufacturing system. A description of all problems involved with the design process is presented.

The second part analyzes the design problems and shows a planning concept, starting with chapter four which analyzes the environment of the design process. The preceding, parallel and follow-up planning activities enable possible objectives, restrictions and decision variables for the design process to be deduced. After the problems of the design have been investigated and the environment of the design process examined, the next step is to outline a concept for the design process itself. This is done in chapter five. The planning process will be divided into two planning stages. The basic planning stage comprises the selection of the production system(s) and the selection of the major equipment of each system. The second stage is given over to more detailed planning. Here layout decisions and system implementation are considered.

The major part of this thesis, however, comprises the third part, which is devoted to the tools and models for the design. It begins in chapter six with a presentation of the applicable tools. Therein general tools are described, followed by domain specific tools, which are specially designed to deal with specific problem areas. Some tools are presented in more detail to allow a better understanding of the solution procedures for the models which follow later. In chapter seven some important aspects of modelling the basic planning stage are outlined. The dominant role of mathematical programming is explained here. Also the relationship between dynamic and static system behavior modelling is analyzed. Dynamic system behavior modelling considers, in contrast to static system behavior modelling, nonlinearities, primarily caused by queueing processes in front of machines. Static modelling neglects these aspects, but allows, through linearization, an easier evaluation of design aspects. Closely related to this issue is the consideration of flexibility aspects during modelling. Depending on flexibility requirements it is shown which is appropriate, - deterministic or stochastic modelling. Furthermore a classification scheme is given according to the decision variables used within each model and finally a descriptive cost model is shown which lists all the major cost factors to be considered during the design process. In chapter eight optimization models for the design of flexible manufacturing systems are presented. It is argued that existing models lack some important features. Based on this observation, new models with their solution procedures are presented. They allow more appropriate cost structures, based on the actual arrangements of components to be considered. Unlike previously published models, which consider only a fixed production demand, changing levels in production requirements are also taken into consideration. However, the major objective is to provide some models for equipment selection reflecting dynamic system behavior. In the literature so far, there have been no models available for this model class. Finally in the last chapter some conclusions and recommendations for the design are given and directions for further research are indicated.

2. Flexible manufacturing systems

A flexible manufacturing system can be defined as *a computer controlled production system capable of processing a variety of part types*.[3]

Below a description of the components of flexible systems found in manufacturing is given. Subsequently the different classes of flexible systems which can be configured by these components are defined. Afterwards definitions of different part sets in manufacturing are specified. Finally a short definition of flexibility follows and a classification scheme for flexibilities of flexible manufacturing systems is presented.

2.1 Components and classes of flexible systems

The major components of a flexible manufacturing system are given by (fig. 2.1):

- CNC-machines and load/unload stations,
- transportation systems for parts and tools,
- and a computerized planning and control system.

---

3   Van Looveren, A.J., Gelders, L.F., Van Wassenhove, L.N.: A Review of FMS Planning Models, in: Modelling and Design of Flexible Manufacturing Systems, Ed.: A. Kusiak, Amsterdam 1986, p.3

fig. 2.1: A small flexible manufacturing system consisting of two machine tools[4]

In most installations of flexible manufacturing systems an incoming raw workpiece is first fixed onto a pallet at a load/unload station. Then it is moved via a material handling system to queues at the CNC-machines, where it is processed. If the required machine is empty, the pallet is directly loaded onto the machine. Otherwise the pallet is stored at a local buffer. When a machine finishes the process the part leaves the machine and the next part, if available, enters. If a further, different processing of the leaving part is necessary, the material handling system transports the pallet to the next machine. If, however, processing for that part is finished, it leaves the system via the load/unload station.

The part flow performed by the material handling system is completely directed by the control computer, which acts as a traffic coordinator. The material handling system itself can consist of carousels, conveyors, carts, robots, or a combination of these.

Different CNC-machines, for example, machining centers, drilling machines, vertical turret lathes, washing machines or inspection machines can be found in flexible manufacturing systems. By means of local tool magazines and automatic tool handling systems, tools for metal-removing machines can be quickly changed and therefore different part types easily processed.

Based on the arrangement of CNC-machines and material handling systems the following classification scheme for flexible production systems can be obtained. The following classes can be distinguished (fig. 2.2):[5]

Flexible manufacturing cell: A flexible manufacturing cell consists of several CNC-

---

4   NN: Ihre verantwortlichen Partner für erfolgreiche Fertigungsstätten, prospectus from Dörries Scharmann GmbH, Düren, Mönchengladbach 1988

5   for alternative classification schemes see for example: Browne, J., Dubois, D., Rathmill, K., Sethi, S.P., Stecke, K.E.: Classification of flexible manufacturing systems, in: The FMS Magazine, April 1984, pp.114-117; Kusiak, A.: Flexible Manufacturing Systems: a structural approach, in: IJPR, 23(1985)6, p.1058; Erkes,

machines augmented by part buffers, tool changers and pallet changers. The CNC-machines themselves can be chosen according to process (identical machines) or to product type (not necessarily identical machines).[6] Unless otherwise stated, the analysis in the following chapters is restricted to cells consisting of identical machines. If the cell consists only of one CNC-machine, it is also referred to as being a flexible manufacturing module or a flexible machining cell.[7]

Flexible machining system: A flexible machining system consists of a collection of flexible manufacturing cells connected by an automated transportation system.[8] There is no restriction on the part flow, i.e. each part can go from one cell to any other cell. To ease modelling for further analysis, a collection of one or more identical load/unload stations incorporated in a flexible machining system is here also defined as a cell.

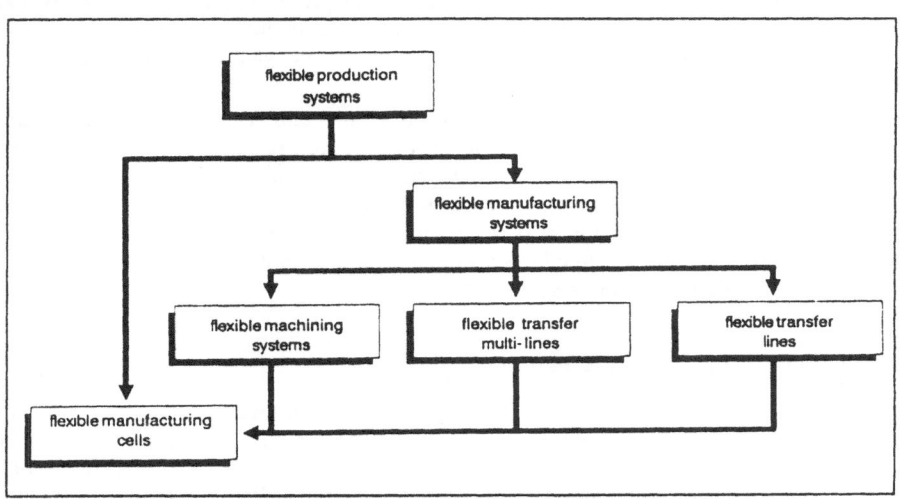

fig. 2.2: Classification of flexible production systems in manufacturing

Flexible transfer line: In flexible transfer lines machine tools are arranged according to the flow shop principle. The flow of different part types through the system follows a given single route.[9]

Flexible transfer multi-line: A flexible transfer multi-line consists of several interconnected flexible transfer lines. It can be considered as a mixture of a pure flexible machining system and a flexible transfer line.

K., Schmidt, H.: Flexible Fertigung, in: VDI-Z, 126(1984)15/16, pp.577-591

6　Warnecke, H. J., Steinhilper, R.: Flexible manufacturing systems; new concepts; EDP-supported planning; application examples, in: Proc. 1st Intern. Conf. on Manufacturing Systems, Brighton U.K. 1982, p.354

7　Dupont-Gatelmand, K.: A survey of flexible manufacturing systems, in: J. Manuf. Syst., 1(1981)1, p.4; Browne, J., Dubois, D., Rathmill, K., Sethi, S.P., Stecke, K.E.: Classification of flexible manufacturing systems, in: The FMS Magazine, April 1984, p.116

8　Browne, J., Dubois, D., Rathmill, K., Sethi, S.P., Stecke, K.E.: Classification of flexible manufacturing systems, in: The FMS Magazine, April 1984, p.116

9　Browne, J., Dubois, D., Rathmill, K., Sethi, S.P., Stecke, K.E.: Classification of flexible manufacturing systems, in: The FMS Magazine, April 1984, p.116

Flexible manufacturing systems: All systems, which are somewhere in between the two extremes of flexible machining systems and flexible transfer lines are considered as flexible manufacturing systems.[10]

2.2 Part sets in flexible manufacturing

The parts produced on a system can be structured in different sets.

Part family: The set of parts, which are produced on a system over a longer time horizon, is defined as its part family. Its qualitative structure is given by the part types it comprises, its quantitative structure by the number of parts produced.

Part mix: A part mix is given by the parts simultaneously produced on a system, i.e. over an infinitesimally small time horizon. As for part families, its qualitative structure is given by the part types it comprises, its quantitative structure by the number of parts produced simultaneously.

Part group: If a subset of part types is produced on the same (alternative) route(s) and with almost identical workloads at each cell (for each route), this subset is considered as a part group. Here a route is characterized by the set of required workloads at the cells visited necessary to perform the required operations for the part group. Thus, if several alternative sets are possible for a part group, alternative routes are available. As a consequence a flexible transfer line has only one part group, and the part group is identical with the qualitative structure of its part family.

2.3 Flexibility in flexible manufacturing

Flexibility is often referred to as an advantage of flexible manufacturing systems. This chapter first presents some general definitions of flexibility to give some insight into its properties. However, to include flexibility considerations into the design process of a flexible manufacturing system more specific definitions of the different kinds of flexibilities found in these systems are necessary. Therefore the explanations are supplemented by a special classification scheme for flexibilities found in flexible manufacturing systems.

For a general definition of flexibility the one given by Mandelbaum is taken. He defines flexibility as the ability to respond effectively to changing circumstances.[11] This definition implies that there is an *object* which is able to respond, and *goals* which allow the effectiveness of the response to be judged. By examining these two, further distinctions can be made.

For an object confronted with changing circumstances, two ways of responding are possible. The first one is that the object simply is capable of continuing functioning effectively despite the change. This is defined as *state* flexibility. The second way of responding is, that further actions are necessary to allow the object to meet new

10   A very similar definition for flexible manufacturing systems, which includes further flexible machining cells, can be found in Browne, J., Dubois, D., Rathmill, K., Sethi, S.P., Stecke, K.E.: Classification of flexible manufacturing systems, in: The FMS Magazine, April 1984, p.116

11   Mandelbaum, M.: Flexibility in decision making: an exploration and unification, Ph.D. Thesis, Dept. of Industrial Engineering, University of Toronto, Ont., 1978, p.20

circumstances. This is referred to as *action* flexibility.[12] Both, state and action flexibility are complementary. If an object has a high degree of state flexibility, it requires less or no action flexibility for changing circumstances. On the other hand, if only a low degree of state flexibility is available, more action flexibility might be necessary to meet the circumstances. If cost considerations are neglected, state flexibility is, in general, more desirable than action flexibility. This is due to the fact that state flexibility responds immediately, whereas action flexibility first requires some kind of action and, depending on the kind of action, a considerable amount of time might be required until the new circumstances are met.

It should be noted that the above distinction has its limitations. In a broad sense every change is related to some kind of action, hence state flexibility could also be considered as necessitating some action. For example a CNC-machine which can produce two different part types might be considered as being state flexible for these part types. However, to switch from one part type to another, some action is required, e.g. the change of the CNC-program. Therefore, only those actions are considered which either consume a lot of time or produce non-negligible costs. Thus action flexibility only relates to actions of this type.

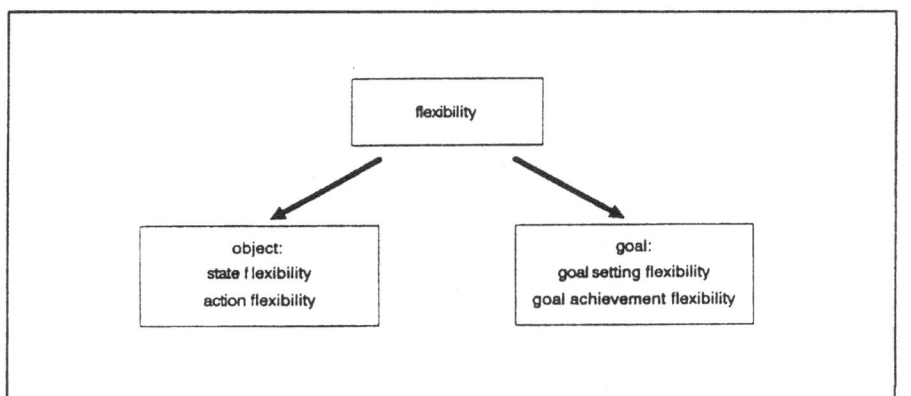

fig. 2.3: Flexibility and its definitions

Another distinction of flexibilities in conjunction with goals is done by Meffert.[13] He defines a flexibility concerning the setting of objectives itself "Zielsetzungsflexibilität" (goal-setting flexibility), and a flexibility concerning the way of achieving these objectives "Zielerreichungsflexibilität" (goal-achievement flexibility). Fig. 2.3 summarizes once more these general aspects of flexibility.

Having explained the characteristics of flexibility, its advantages are clear. Now the question arises of how to provide enough flexibility. Unfortunately the supply of flexibility is not unlimited, because flexibility causes costs.[14] This fact means that flexibility must be

---

12   Mandelbaum, M.: Flexibility in decision making: an exploration and unification, Ph.D. Thesis, Dept. of Industrial Engineering, University of Toronto, Ont., 1978, p.20; Jacob uses the expressions "Bestands- und Entwicklungsflexibilität", see Jacob, H.: Unsicherheit und Flexibilität: Zur Theorie der Planung bei Unsicherheit, in: ZfB, 44(1974)5, p.322

13   Meffert,H.: Zum Problem der betriebswirtschaftlichen Flexibilität, in: ZfB, 39(1969),p.787

14   Reichwald, R., Behrbohm, P.: Flexibilität als Eigenschaft produktionswirtschaftlicher Systeme, in: ZfB,

planned, i.e. the optimal supply of flexibility must be found. In order to do this, it is necessary to forecast the demand for flexibility. Different evaluations must be made depending on the stochastic nature of the demand, i.e. if it is a demand for flexibility under

- certainty

- risk

- uncertainty.

The following concentrates on demand under certainty. However, it must be pointed out that flexibility is often considered as the ability to react efficiently to *unforeseen* changing circumstances, and its demand therefore is uncertain.

In order to plan the demand for flexibility of a flexible manufacturing system it is necessary to specify further the above abstract definitions of flexibility and quantify them. In deciding which aspects and parameters are relevant there is some loss of generality. Depending on which aspects are emphasized, a vast amount of classification schemes for flexibilities in flexible manufacturing can be found.[15] The following scheme given by Browne et al. can be considered as one of the most comprehensive ones.[16] They distinguish between:

Machine flexibility: Ease (time required) of making the changes necessary to process a given part mix (action flexibility).

Operation flexibility: Ability to interchange ordering of operations on a part (state flexibility).

Routing flexibility: Ability to process a given part mix on alternative machines (state flexibility).

Process flexibility: Ability to produce a given part mix (state flexibility).

Product flexibility: Ease (time required) to change from one part mix to an other (action flexibility).

Volume flexibility: Ability to operate profitably at varying overall levels (state flexibility).

Expansion flexibility: Ease (time required) to add capacity and capability (action flexibility).

Production flexibility: The universe of part types that can be processed (state flexibility).

Disregarding the problem of quantifying these flexibilities exactly,[17] it can be stated that

---

53(1983)9, p.841

15   Browne, J., Dubois, D., Rathmill, K., Sethi, S.P., Stecke, K.E.: Classification of flexible manufacturing systems, in: The FMS Magazine, April 1984, pp.114-117; Buzacott, J. A.: The fundamental principles of flexibility in manufacturing systems, in: Proc. of the 1st Intern. Conf. on Flexible Manufacturing Systems, Brighton U.K. 1982, pp.13-22; Carter, M.F.: Designing flexibility into automated manufacturing systems, in: Proc. 2nd ORSA/TIMS Conf. on Flexible Manufacturing Systems: Operations Research Models and Applications, Ed.: Stecke, K.E., Suri, R., Amsterdam 1986, pp.107-118; Falkner, C.H.: Flexibility in manufacturing plants, in: Proc. 2nd ORSA/TIMS Conf. on Flexible Manufacturing Systems: Operations Research Models and Applications, Ed.: Stecke, K.E., Suri, R., Amsterdam 1986, pp.95-106; Gerwin, D.: An agenda for research on the flexibility of manufacturing processes, in: IJOPM, 7(1987)1, pp.38-49; Maier, K.: Die Flexibilität betrieblicher Leistungsprozesse, Frankfurt am Main 1982; Schaefer, F.-W.: System zur Planung und Nutzung der Flexibilität in der Fertigung, Diss. der TH-Aachen, Aachen 1980; Zelenovic, D.M.: Flexibility - a condition for effective production systems, in: IJPR, 20(1982)3, pp.319-337

16   Browne, J., Dubois, D., Rathmill, K., Sethi, S.P., Stecke, K.E.: Classification of flexible manufacturing systems, in: The FMS Magazine, April 1984, pp.114-117

17   The author is grateful to Dr. Kathryn E. Stecke for drawing his attention to these problems, which have yet

the demand for each type of flexibility can be described by considering that our system needs to fulfill a certain goal, i.e. certain production requirements. From these requirements demands on flexibilities can be derived by examining different time horizons. If a long time horizon is observed, the production demand is given by the part family to be produced on the system. With the qualitative structure of the part family the universe of part types that must be processed, i.e. the production flexibility, is determined. Furthermore the quantitative structure of the part family requires a certain amount of volume and expansion flexibility of the system. If a short time horizon or real time requirements are studied, the necessary demand for flexibility depends on the actual part mix we want to produce in the system. By definition, requirements in machine, process, product, routing and operation flexibility depend on the part mix.

demand source	flexibilities
part family	volume expansion production
part mix	machine process product routing operation

fig. 2.4: Flexibilities and their demand source

3. Planning problems of flexible manufacturing systems

According to Kusiak, the planning problems of flexible manufacturing systems can be divided into two groups, i.e. in design and operational problems.[18] Whereas in the first group the selection of equipment and considerations about the system's layout are relevant, the second is concerned with the optimal utilization of an already existing system.

In the following a short summary of both groups is given. Because this thesis emphasizes design problems, operational problems will be described here only for completeness. Therefore the latter are presented here in a structured framework, whereas the design problems are structured and analyzed later in more detail.

3.1 Design problems

If the management considers implementing a flexible manufacturing system, the question of the system's specification arises.[19] Here quite a few problems need to be

to be solved satisfactorily.

18   Kusiak, A.: Flexible Manufacturing Systems: a structural approach, in: IJPR, 23(1985)6, p.1063

19   for the discussion of design problems see also: Stecke, K.E.:Design, Planning, Scheduling, and Control Problems of Flexible Manufacturing Systems, in: Annals of OR, 3(1985)3, p.4; Van Looveren, A.J., Gelders, L.F., Van Wassenhove, L.N.: A Review of FMS Planning Models, in: Modelling and Design of

solved:

One problem arises with the specification of the part spectrum to be manufactured on the system. Therefore it is necessary to generate a part family (or families) for the system(s). This is done by analyzing the whole spectrum of parts under technical and economic criteria. To include dynamic aspects of changing products and demands, a reasonably long-term production plan should support this selection process. "Ideally, one should make a specification of what will be manufactured by the system during its whole lifetime."[20]

Another problem is the specification of the machines to be included. The number of necessary machines of each machine type needs to be specified. These decisions can be made based on the process plan, or, if available, on several alternative process plans for each part.

Furthermore the material handling systems and their capacities need to be specified.

An analysis must be performed, which reveals the required flexibilities of the system.

The number and kinds of pallets in the system and the necessary fixtures for the parts have to be specified.

An important role is played by the computer system with its hard- and software. The hierarchy and architecture of the computer network has to be considered as well as the planning and control strategies to be used.

Furthermore a problem arises in the evaluation of local buffers needed at every cell and the size of the central buffer. On the one hand, buffer space should be large enough to avoid congestions, i.e. blocking, in the system, but on the other hand, it should not be so large as to cause excessive use of space and high buffer costs.

Finally layout considerations and the integration of all the above-mentioned equipment have to be taken into account.

3.2 Operational problems

The operational problems relate to a flexible manufacturing system which is already configured and integrated into the production environment. Hence the flexible manufacturing system is part of the whole multi-stage production. The general problem now is to organize production so that the master production plan is satisfied and the system resources are used in an efficient way.

There exist a number of hierarchical classification schemes for operational problems in flexible manufacturing which differ only in their details.[21] In the following it is referred to

Flexible Manufacturing Systems, Ed.: A. Kusiak, Amsterdam 1986, p.7; Kusiak, A.: Flexible Manufacturing Systems: a structural approach, in: IJPR, 23(1985)6, p.1063

20   Van Looveren, A.J., Gelders, L.F., Van Wassenhove, L.N.: A Review of FMS Planning Models, in: Modelling and Design of Flexible Manufacturing Systems, Ed.: A. Kusiak, Amsterdam 1986, p.6

21   Stecke, K.E.:Design, Planning, Scheduling, and Control Problems of Flexible Manufacturing Systems, in: Annals of OR, 3(1985)3, p.4; Van Looveren, A.J., Gelders, L.F., Van Wassenhove, L.N.: A Review of FMS Planning Models, in: Modelling and Design of Flexible Manufacturing Systems, Ed.: A. Kusiak, Amsterdam 1986, p.7; Kusiak, A.: Flexible Manufacturing Systems: a structural approach, in: IJPR, 23(1985)6; p.1063; Kiran, A.S., Tansel, B.C.: The System Setup in FMS: Concepts an Formulation, in: Proceedings of the Second ORSA/TIMS Conference on Flexible Manufacturing Systems, Ed.: K.E. Stecke, R. Suri, Amsterdam 1986, pp.321-332; Suri, R., Whitney, C.: Decision Support Requirements in Flexible

an hierarchical system of two levels - the tactical and the operative level.[22] The main difference between the levels is to be seen in the different time horizon used.

On the tactical level the flexible manufacturing system is first considered as one unit. To begin with, the first problem is to allocate (allocation problem) the production orders for parts listed in the master production schedule to alternative production strategies, i.e. to the job shop, flow shop or the flexible manufacturing system, if available.

If parts are assigned to the flexible manufacturing system, they now have to be sequenced into batches (batching problem). The objective here is to organize production in such a way that orders are completed on time, taking into account limitations and restrictions of the available resources. Note that because of machine flexibility and the fact, that the set-up of parts can be done while the machine is still working on other parts, a batch for a flexible manufacturing system can consist of different part types.

Having first considered the system on the tactical level as a whole, the next step is to consider the operational aspects in between the components of the system itself, i.e. the decisions on how to manufacture the parts and which operation should be performed on which machine are examined. Therefore a decision about the loading of each machine is to be made (loading problem) under the constraint of limited resources available. Limited resources can be for example machines, pallets, fixtures and tools.

When the batches are finally created and the loads assigned, the necessary preparations for production can be performed in time.

On the operative level, problems concerning the real time operation of the system are relevant. Here, as on the tactical level, two different viewpoints can be distinguished, regarding, on the one hand, the system as one unit, and, on the other, as a collection of single components.

The former considers the flow of pieces into the system and is therefore referred to as the release problem. With the help of the planning results of the tactical level and the current status of the system, decisions about the input to the system are made. The current system status in this respect comprises data about machine breakdowns, types of pallets and fixtures available, deviations from the desired production rate etc..[23]

Manufacturing, in: J. Manuf. Syst., 3(1984)1, pp.61-69

22   with minor changes it is based on the work of Van Looveren et al.: Van Looveren, A.J., Gelders, L.F., Van Wassenhove, L.N.: A Review of FMS Planning Models, in: Modelling and Design of Flexible Manufacturing Systems, Ed.: A. Kusiak, Amsterdam 1986, p.7

23   Van Looveren, A.J., Gelders, L.F., Van Wassenhove, L.N.: A Review of FMS Planning Models, in: Modelling and Design of Flexible Manufacturing Systems, Ed.: A. Kusiak, Amsterdam 1986, p.7

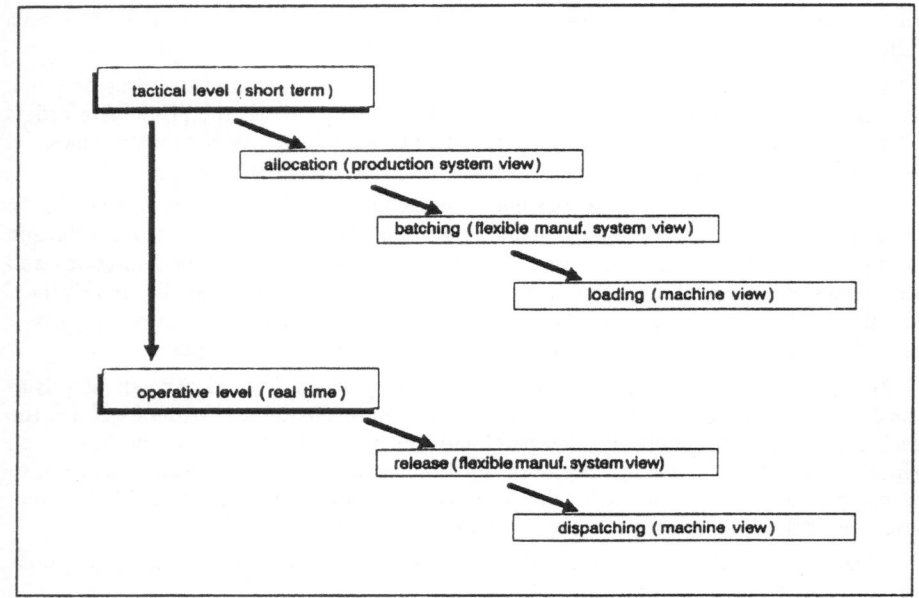

fig. 3.1: Operational problems

The second viewpoint considers the flow in the system and is referred to as the dispatching problem. Here decisions must be made about the control of the part flow. (Note that for flexible transfer lines the part flow is given.) Priorities have to be assigned to the parts in the queue in front of each station. The assignment of buffer space has to be done in such a way that blocking is avoided, and in case of machine breakdowns, special strategies are needed to allow a smooth continuation of production.

PART B: ANALYZING THE DESIGN PROCESS

4. The design process

In the previous chapters an introduction to the characteristics and planning problems of flexible manufacturing systems has been given. The objective of this chapter is to examine the structure of the design process in the context of the other surrounding planning processes. Therefore first the environment of the design process, i.e. the preceding, parallel and follow-up planning processes, are studied and their interactions with the design process examined. Based on these considerations, conclusions about the objectives, decision variables and restrictions of the design process can be drawn.

4.1 The environment of the design process

To analyze the environment of the design process, first a closer look at the preceding and parallel planning processes, i.e. finance, investment, capacity and long-range production planning is taken. Then the follow-up operational planning process is studied.

4.1.1 Preceding and parallel planning processes

The decision to invest in a flexible manufacturing system can be regarded as a strategic decision. This is due to its long term impact on the company's policy caused by high investment costs (between 5-30 million DM) and the long lifetime of such a system (in general between 8-10 years).[24]

Based on a long-range production plan, the required capacity for production has to be determined.[25] This is done by comparing the necessary against the available amount of capacities. If there is a capacity gap for some parts, i.e. the available capacity is less than needed, the problem arises of how to produce those parts if they are not to be bought externally. Therefore a selection between different possible types of production systems must be performed.[26] Depending on technological and economic criteria, the management can in general chose between pure flow shop production, job shop production or flexible manufacturing systems. This has to be done within the available budget, obtained from the long-range finance plan of the business.[27]

---

24 Wildemann, H.: Investitionsplanung und Wirtschaftlichkeitsrechnung für flexible Fertigungssysteme (FFS), Stuttgart 1987, p.138

25 Hahn, D.: Planung- und Kontrollrechnung, PuK, Wiesbaden 1974, p.64-67; Arbeitskreis "Langfristige Unternehmensplanung" der Schmalenbach Gesellschaft: Strategische Planung, in: Planung und Kontrolle, editor: H. Steinmann, München 1981, pp.23-45

26 for a general classification of different production systems see: Zäpfel, G.: Produktionswirtschaft, Berlin, New York 1982, pp.15-20; Schmitt, T.G., Klastorin, T., Shtub, A.: Production classification system: concepts, models and strategies, in: IJPR, 23(1985)3, pp.563-578

27 Hahn, D.: Planung- und Kontrollrechnung, PuK, Wiesbaden 1974, p.64-67

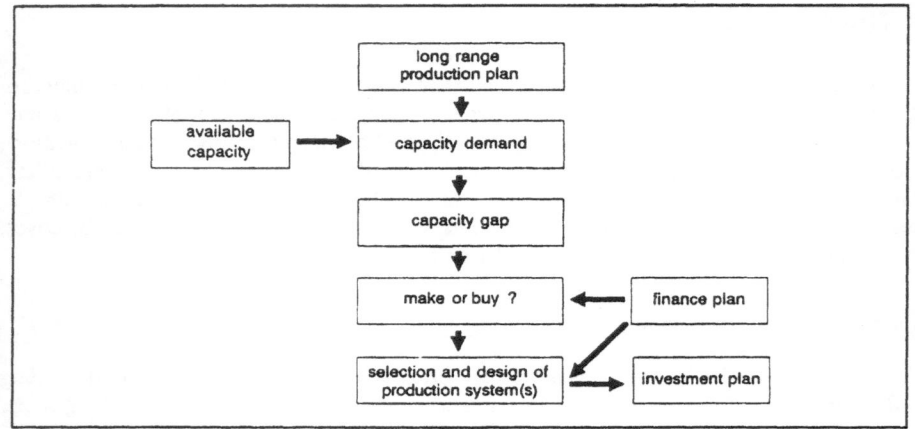

fig. 4.1: Preceding planning processes

To support this decision process, alternative scenarios are generated and alternative designs developed and evaluated.[28] This point can be seen as the first step in the design process. For a given part spectrum with given production requirements, decisions must be made, taking into account the following questions:

- what kind of production system(s) to use and

- how the production system(s) are realized.

Complications for these decisions arise through the fact that the two above-mentioned questions are strongly interrelated, and that the part spectrum and production requirements can be of a changing nature, i.e. stochastic and dynamic aspects have to be considered. Furthermore these decisions must also fit in with the long-range investment- and finance plan (see fig.4.1).

4.1.2 Follow-up planning process

The interface between the design process and the follow-up operational planning process can be seen in the availability of the necessary production capacity, i.e. a flexible manufacturing system or other alternative production systems. The design process ends when the production system is implemented and manufacturing begins. Thus we are now confronted with operational problems. Design problems can arise again if new capacities are required and thus new production systems or system extensions become necessary.

---

28   Adams, F.P., Cox, J.F.: Manufacturing Resource Planning: An Information Systems Model, in: Long Range Planning, 18(1985)2, pp.86-92

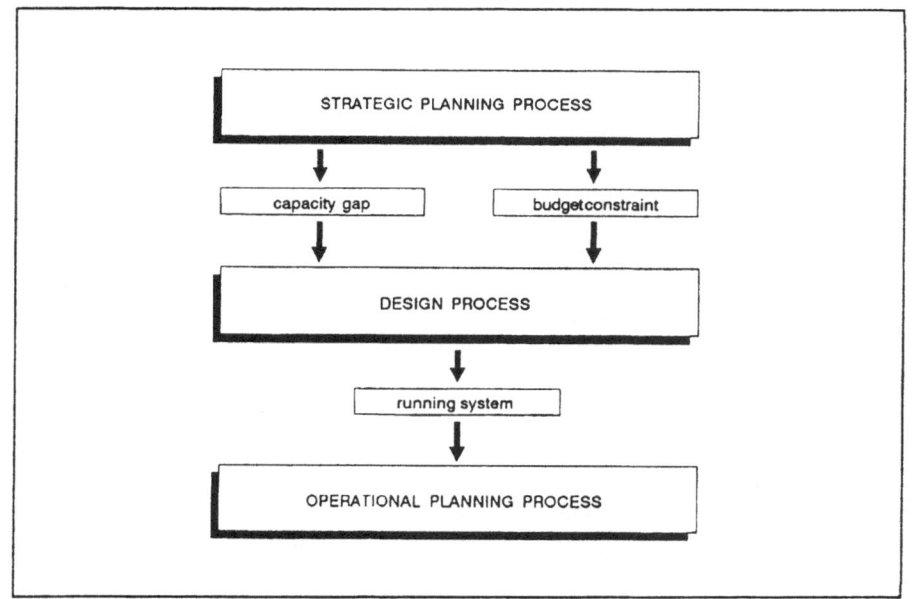

fig. 4.2: The environment of the design process

4.2 Decision variables, objectives and restrictions of the design process

Based on the environmental analysis of the design process in the preceding chapter (see fig.4.2), the decision processes of the design process itself are now analyzed. A detailed examination concentrating on flexible manufacturing systems follows. However, only minor changes are necessary to apply these results to other production systems.

4.2.1 Decisions

Decisions and their related variables can be obtained by examining the planning process which follows the design process, i.e. the operational planning process. As stated in chapter 4.1.2, the latter assumes a production system already in operation having the necessary capacities for production. Therefore the design process must incorporate all those decisions, which lead to such an operational system. They comprise:[29]

- Selection of production system(s) used to produce the required products;

- Selection of parts and the amounts of them to be produced on a flexible manufacturing system over the whole lifetime of the system;

---

29   for very similar specifications see Kalkunte, M.V., Sarin, S.C., Wilhelm, W.E.: Flexible Manufacturing Systems: A Review of Modelling Approaches for Design, Justification and Operation, in: Flexible Manufacturing Systems: Methods and Studies, Ed.: A. Kusiak, Amsterdam 1986, p.7; Kusiak, A.: Flexible Manufacturing Systems: a structural approach, in: IJPR, 23(1985)6, p.1063; Stecke, K.E.:Design, Planning, Scheduling, and Control Problems of Flexible Manufacturing Systems, in: Annals of OR, 3(1985)3, p.4-6

- Selection of the equipment for the production system. For a flexible manufacturing system this includes:

  Selection of the CNC-machines, i.e. their type and the number of each;

  Defining the number of load/unload stations in the system;

  Selection of the transportation, the material handling and tool handling systems;

  Defining the number of buffers and the inventory system;

  Defining the number of pallets and fixtures needed;

- Defining the type and structure of the planning and control system;
- Defining the layout of the system;
- Defining the number and skills of personnel needed;
- Defining the demand for and scope of each type of flexibility.

It should be noted that these decisions are strongly interrelated. This can be easily demonstrated by a simple model of causal relationships between the beginning of the design process, i.e. the given capacity gap for some parts, and the result of the design process, i.e. the installed production system(s).

Our model consists of a causal chain which depicts the path of relationships to obtain one single alternative for the assignment of a part to a production system. First for the part its (alternative) process plan(s) have to be examined. Then each process plan is taken and further subdivided into single operations. For each operation a specific set of tools is necessary. To perform the operation, however, a machine tool is also required. Just to obtain a *feasible* solution for our part under consideration, the machine tools needed to produce the part must be supported by all the other supplementary resources (for flexible manufacturing systems see list above) to obtain an operational system. To obtain a good and not only a feasible solution for a part, the interdependences on other parts and their causal chains must be taken into consideration. This is due to the fact that, as soon as tools, machine tools or other pieces of equipment are selected for the system, there are consequences for other parts. This is shown in fig.4.3 by horizontal arrows between tools, machine tools and the equipment.

From this model of causal links between the part and the resulting production system it can be concluded by observing the length of the chain that we are faced with an overwhelming array of possibilities, even if only a few alternatives exist at each link of the chain and interdependences to other parts are neglected.

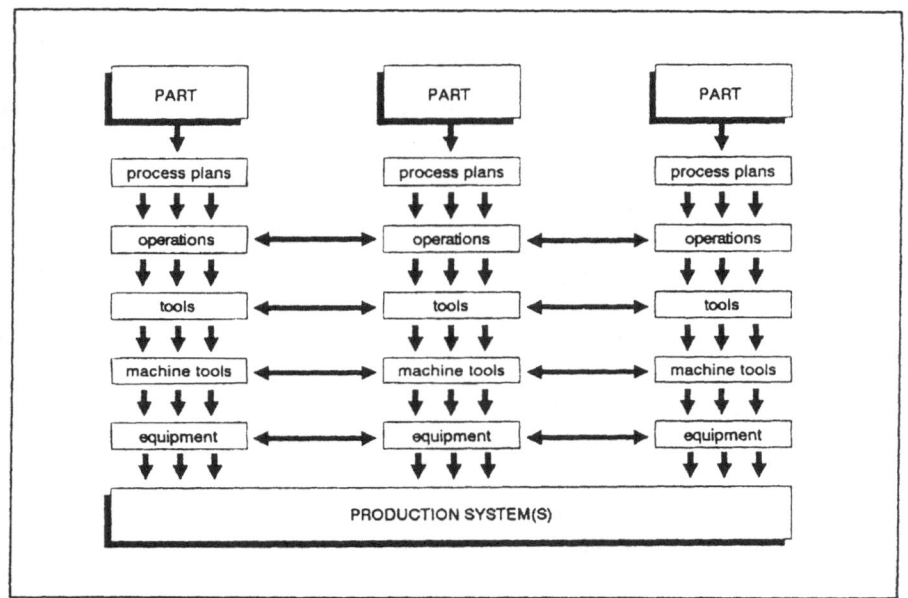

fig. 4.3: Model of causal relationships of the design of production systems

4.2.2 Objectives

In the previous chapter decisions necessary for the design process were discussed. Now the question arises as to which criteria these decisions should be based on, i.e. according to which objectives should a good or optimal solution be selected? Kalkunte et al., for example, name three different objectives: cost, productivity and product quality.[30]

In the following the analysis is restricted to productivity. Productivity can be described as a relationship between input and output of production and thus allows its economic feasibility to be measured.[31]

In order to define input and output of a production system such as a flexible manufacturing system, those decision variables which have an effect on the input or output of the system must be considered. "Effective productivity measurement requires the development of an index that identifies the contribution of each factor of production and

30   Kalkunte, M.V., Sarin, S.C., Wilhelm, W.E.: Flexible Manufacturing Systems: A Review of Modelling Approaches for Design, Justification and Operation, in: Flexible Manufacturing Systems: Methods and Studies, Ed.: A. Kusiak, Amsterdam 1986, p.6

31   Adler, P.S.: A Plant Productivity Measure for "High-Tech" Manufacturing, in: Interfaces, 17(1987)6, p.82; Chew, B.W.: No-Nonsens Guide to Measuring Productivity, in: HBR, (1988)1, p.114; It should be noted, that in german literature for productivity, i.e. "Produktivität", several definitions can be found. In essence however, it always revolves around a relationship between input and output. See for example: Zimmermann,G.: Ergiebigkeitsmaße der Produktion, in: Handwörterbuch der Produktionswirtschaft, Ed.: W. Kern, Stuttgart 1979, pp.520-528; Lehmann, M.R.: Wirtschaftlichkeit, Produktivität und Rentabilität (I), in: ZfB, 28(1958), pp.537-557,614-620;

then tracks and combines them."[32] Based on the analysis in the previous chapter one can derive as output the parts which are produced on the system, and as input all the resources necessary to produce these parts, e.g. the equipment for the system, the personnel, raw material etc.. To allow a comparison of differently measured inputs and outputs, the transformation to a common unit of measure is necessary. This can be accomplished by measuring in monetary units.

Based on the definition of productivity three objectives can be derived, where each of them can be chosen depending on the circumstances surrounding the design process:

- Minimize only input (costs, present value of cash outflows), while output is constant.
- Maximize output (part production), while input is constant.
- Optimize the relation between input and output.

To decide which of the three above-mentioned ways is used to optimize the system during the design process, the results of the preceding planning processes have to be considered. As already described in chapter 4.1.1, the planner might wish to configure the production system(s) in such a way, that a given capacity gap is closed. Consequently he chooses the first objective. However he might discover that the solution found consists of a flexible manufacturing system which causes initial investment costs exceeding the available budget. Alternatively the planner might then either try to make adjustments on the preceding planning processes, i.e. the budget, or he might try to find a solution by using the second objective and produce as much as possible within the available budget. If, however, there is neither a limited budget nor a restriction on output, the last objective is applicable.

4.2.3 Restrictions

Some restrictions for the design process can easily be derived from the environment of the design process described in chapter 4.1. From the long-range production plan a capacity gap is given, and from the finance plan the available budget to fill this gap (see fig.4.2) is obtained.

Further restrictions are given by:

- technological constraints due to the physical properties of the parts to be produced,
- technological constraints due to the properties of the potential equipment,
- limited space for the layout of the system,
- necessary throughput times because of customer service requirements or because of synchronization necessary with other down- or upstream production processes.

5. A concept for the design process

In the previous chapters first the design problems and then the environment of design process were studied. From this the structures of the design process, i.e. its necessary decisions, objectives and restrictions were obtained.

---

32   Chew, B.W.: No-Nonsens Guide to Measuring Productivity, in: HBR, (1988)1, p.114

Based on these results, our principal aim is now to construct a decision model which enables us to solve the design problems. This could be done within a single model incorporating all the causal relationships between the decision variables in an adequate way, while considering all the given restrictions. However, for three reasons it seems to be questionable whether such a simultaneous procedure is applicable.

-   Complexity problem: As already shown in chapter 4.2.1 by a simple model of causal relationships, the design problems are very complex and the set of alternatives grows rapidly with the number of decision variables.

-   Method problem: It seems to be questionable whether the available tools are powerful enough to construct and solve a satisfactory homomorphic model of reality.

-   Data problem: To be able to explore all alternatives a lot of data is required.

As a possible way to circumvent these problems further structuring of the design process into a sequential procedure is proposed. This allows the reduction of the problem to a set of easier subproblems, which are simpler to solve and necessitate less data. However two disadvantages are bound up with this concept:

-   Firstly, it is possible for good or even optimal solutions to be discarded at the first steps of the sequential procedure, because they do not yet appear promising.

-   Secondly, with the solution obtained from the first steps it might not be possible to generate a feasible solution in the following steps.

Nevertheless, these disadvantages can be avoided if the procedure is structured and modeled in such a way, that the neglected causal relationships between decision variables of different subproblems are of minor importance. This can be achieved, for example, by performing first those decisions which have a major impact on the planning objective. Moreover the sequential planning procedure can be performed repeatedly with a feedback to the initial planning steps to ensure the generation of a feasible solution.

For the design of production systems a two-step planning procedure is suggested. The separation of decision variables is done by selecting first those variables which have a major influence on the objective and allow system costs or cash outflows for the system (input) and system behavior (output) to be estimated as accurately as possible during the first planning step. This allows coordination with other planning areas, i.e. the investment plan and the long-range finance plan. Consequently the first step consists of basic decisions including the choice of production system(s), and the selection of the major pieces of equipment for each system. The second step comprises a more detailed planning stage. Here the final layout and the timing for the implementation of the system is considered. The following picture summarizes the concept once more (see fig.5.1). In the next chapters both planning stages are described. However, the emphasis here is on the basic planning stage.

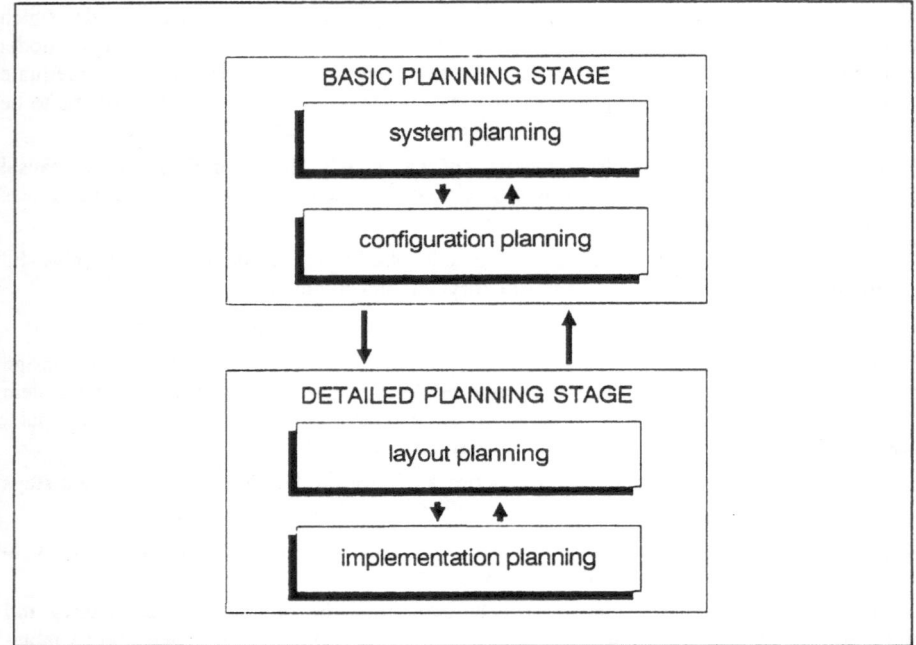

fig. 5.1: A planning concept for the design

5.1 The basic planning stage

Through sequencing the design process only a subset of decision variables is considered, while the objectives remain the same and the number of restrictions might increase. For the decisions during the basic planning stage for *system planning* the following are considered:

-   the part assignment to production systems,
-   the type of production system,

while for the *configuration planning* of a flexible manufacturing system the following decisions are considered:

-   the selection of cells,
-   the number of machines or load/unload stations in each cell,
-   the transportation system and its number of vehicles,
-   the number of pallets and fixtures,
-   flexibility requirements.

However the number of pallets and fixtures might also be subject to the following layout planning phase, depending on how much the planning objective (see chapter 4.2.2 and 4.3) is influenced by it. For example, if costs for pallets and fixtures play a negligible role compared to the other equipment costs, consideration during the layout planning phase might be more appropriate.

Restrictions reduce the set of alternatives during the basic planning stage. Especially during *system selection*, the part assignment to different types of production systems can be simplified by considering further technical and economic restrictions:

-   Technological constraints might permit only a specific type of production system for a part.

-   Sometimes the nature of required operations for a part predestines it for a specific production system. For example, parts with high part set-up times are often advantageous for flexible manufacturing systems. This is due to the fact that through palletizing part clamping can be performed while the machines are still processing other parts.

-   Strategic considerations favour the installation of certain types of production systems.

-   The necessary production and flexibility requirements often reduce the types of production systems under consideration. For example the type of flexible manufacturing system depends on the required machine, process, routing and operation flexibilities according to the part mixes to be produced on the system. If the part mix is fairly constant, consisting of only one or a few part groups a flexible transfer line or multi-line is applicable. On the other hand, if changes in the part mix are irregular and consists of many part groups, a flexible machining system is necessary.

For *configuration planning* the tightness of the given restrictions is determined by flexibility requirements. As shown in chapter 2.3, volume, expansion and production flexibility are mainly determined by the assigned part family. Production flexibility, i.e. the range of part types that the system is able to process, can be derived from an accurate long-term production plan for the system over its lifetime. Volume and expansion flexibility are complementary. Volume flexibility can be considered as the state flexibility and expansion flexibility as the action flexibility for the production volume. Depending on the production forecast, sometimes expansion to a later stage might be better than configuring the system with some overcapacity from the start, so that it can handle all future production requirements. Based on the production forecasts, optimization models must be developed to find the optimal supply for each of the two flexibilities.

The parts that can be *simultaneously* produced on the system, i.e. the part mix, and the capability to change easily to other part mixes, if required, are described by process, machine, product, operation and routing flexibility. Thus, even if all part types can be produced on the system, it does not necessarily follow that they all can be produced simultaneously on the system. This is due to the fact that resource constraints limit this capability. Here especially a limited tool magazine with a limited number of tool slots can restrict the number of part types contained in a part mix. However, for two reasons it is suggested here that tool constraints - and therefore process, machine and product flexibility - should be disregarded during the configuration planning phase:

-   Due to possible technical design features such as an automatic tool handling system with large local and centralized tool storage, tool availability in many cases does not cause such a bottleneck as it sometimes seems.

-   Configuration planning is a long-range planning procedure which puts emphasis on the expected average system behavior. Hence its decisions are based on forecasts of a long-range, less accurate, production plan. Consequently, detailed planning, which includes considerations about the exact timing of when to produce which part mix, cannot be the topic of configuration planning. This subject relates more closely to the operational planning environment.[33]

Routing and operation flexibility enhance the system's capability to deal with unforeseen disturbances, such as machine breakdowns etc.. Furthermore it allows a better balancing of workloads if part mixes are changing. Thus, when the type of production system is already selected, these flexibilities are primarily subject to the layout planning procedure, where control strategies are considered.

5.2 The detailed planning stage

If the basic planning stage is finished, a few problems still have to be solved before the design of the system is completed. If not already done in the configuration planning phase, the number of pallets and fixtures has to be determined. Furthermore the number of buffers and their location must be specified. The structure of the required planning and control system and the amount and skills of personnel needed are additional aspects to solve. The strategies of the control system must be defined. Here operation and routing flexibility are of particular importance. Finally the physical layout, considering given restrictions on space and other resources, must be determined.

After the final physical layout of the system has been established, the planning of the different tasks for the implementation of the system and their timing is made. But not only the physical implementation of the system has to be considered. The necessary organizational adaptations and changes must also be planned.

---

33   see chapter 3.2

PART C: TOOLS AND MODELS

6. Tools for the design process

In this chapter an overview of the categories of tools applicable to the design process is presented. Furthermore, some special tools are described which can be used later to solve the models newly developed in chapter 8.

Classification of these tools can only be very arbitrary. On the one hand, classification based on the more domain-specific character of some tools in contrast to the more general character of other tools is possible. On the other hand, a difference can be made from a more practical user's point of view by distinguishing between two different tool categories according to the *objectives* they pursue. These are evaluative tools, intended to evaluate a given set of decisions, and generative tools, intended to reach a "good" decision. However, as Suri has stated, a tool as for example perturbation analysis does not only evaluate a given set of decisions, but also has some semi-generative capabilities, i.e. it attempts to find directions to improve the existing decision.[34] Furthermore, from a theoretical point of view it is not very satisfactory to consider for example mathematical programming alongside group technology. This is due to the fact that the latter incorporates algorithms from the former in some procedures.[35] Therefore the tools are here classified under the first classification principle.

Under "general tools" those tools are summarized which have a more general character and which are not restricted to a particular problem field. Mathematical programming, queueing networks, Petri nets, computer simulation and perturbation analysis can be grouped here. On the other hand, there are domain-specific tools which are linked to a special field, such as group technology, which is restricted to the structuring of production systems into groups, or knowledge-based systems, which incorporate a domain-specific knowledge base.

6.1 General tools

6.1.1 Mathematical programming

Mathematical programming is concerned with the determination of the optimum of a function of several variables, which also have to satisfy a number of constraints. These constraints can be inequalities and/or equalities.[36]

Depending on the kind of function and constraints, e.g. linear or nonlinear, and the values the variables can have, e.g. continuous or integer, different methods for finding an optimum solution exist. An overview of standard algorithms and methods is given by Eiselt and von Frajer.[37]

---

34   Suri, R.: An Overview of Evaluative Models for Flexible Manufacturing Systems, in: Annals of OR, 3(1985), p.16

35   see for example: Steudel, H.J., Ballakur, A.: A Dynamic Programming Based Heuristic for Machine Grouping in Manufacturing Cell Formation, in: Computers ind. Engng., 12(1987)3, pp.215-222

36   Zoutendijk, G.: Mathematical Programming Methods, Amsterdam, New York, Oxford 1976, p.1

37   Eiselt, H.A., Frajer von, H.: Operations Research Handbook, Standard algorithms and Methods, Berlin, New York 1977

Throughout chapter 8 mathematical programming will play an important role in solving the problems of the design process of flexible manufacturing systems. Some new models will be presented, which are solved by special kinds of mathematical programming tools described in the following chapters.

6.1.1.1 Augmented Lagrangians

In this chapter a general overview of a solution procedure to solve constrained nonlinear programs with the help of Lagrangian functions will be given. A general nonlinear programming problem has the following form:

$$\min f(\underline{x}) \tag{6.1.1.1.1}$$

subject to:

$$c_i(\underline{x}) = 0 \qquad i \in M_e = \{1, \ldots, m_e\}$$
$$c_i(\underline{x}) \leq 0 \qquad i \in M_i = \{m_e + 1, \ldots, m\}$$

In order to simplify the presentation the inequalities are split into a set of binding constraints ($c_i = 0$ with $i = \{m_e + 1, .., m_b\}$) and a set of nonbinding inequalities ($c_i < 0$ with $i = \{m_b + 1, .., m\}$). It is assumed that there is second order differentiability and that the normals of the binding constraints $c_i(\underline{x}^*)$ with $i \in \{1, .., m_b\}$ at the optimal solution $\underline{x}^*$ are linearly independent.

The solution method consists of the conversion of this constrained problem to a sequence of unconstrained minimization problems, having the property that the successive solutions of the unconstrained problems converge to the solution of the constrained problem.

The Lagrangian function of the problem (6.1.1.1.1) is given by:

$$LF(\underline{x}, \underline{\alpha}) = f(\underline{x}) + \sum_{i \in \{1, \ldots, m_b\}} \alpha_i \cdot c_i(\underline{x})$$

for which the Kuhn Tucker first-order necessary conditions yield unique parameters α_i^* such that the equation:[38]

$$\operatorname{grad} LF(\underline{x}^*, \underline{\alpha}^*) = \operatorname{grad} f(\underline{x}^*) + \sum_{i \in \{1, \ldots, m_b\}} \alpha_i^* \cdot \operatorname{grad} c_i(\underline{x}^*) = 0 \tag{6.1.1.1.2}$$

is satisfied. The method first proposed by Powell[39] and Hestenes[40] for equalities and then extended by Rockafellar[41] for inequalities in the constraint set augments the Lagrangian function by penalty terms. In this way constraint violations are incorporated in the Lagrangian function in the form of penalties. Thereby the constrained program can be transformed into an unconstrained augmented Lagrangian penalty function, which in the

38   Fiacco, A.V., McCormick, G.P.: Nonlinear Programming: Sequential Unconstrained Minimization Techniques, New York 1968, p.20

39   Powell, M.J.D.: A method for nonlinear constraints in minimization problems, in: Optimization, Ed.: R. Fletcher, New York 1969, pp.283-298

40   Hestenes, M.R.: Multiplier and Gradient Methods, in: JOTA, 4(1969), pp.303-320

41   Rockafellar, R.T.: A dual approach to solving nonlinear programming problems by unconstrained optimization, in: MP, 5(1973), pp.354-373

formulation of Hestenes and Rockafellar has the following form:[42]

$$LF_a(\underline{x},\underline{\alpha},\underline{w}) = f(\underline{x}) + \sum_{i \in M_e} (\alpha_{1i} \cdot c_i(\underline{x}) + w_1 \cdot c_i(\underline{x})^2)$$

$$+ \sum_{i \in M_i} \left[\begin{array}{ll} w_2 \cdot c_i(\underline{x})^2 + \alpha_{2i} \cdot c_i(\underline{x}) & \text{if } c_i(\underline{x}) > - \alpha_{2i} / (2 \cdot w_2) \\ - \alpha_{2i}^2 / (4 \cdot w_2) & \text{if } c_i(\underline{x}) \le - \alpha_{2i} / (2 \cdot w_2) \end{array} \right.$$

The factor $\underline{w}$ is a weighting factor, which must be chosen large enough. At the optimal solution point $\underline{x}^*$ of the constrained nonlinear program there exists an $\underline{\alpha}$, which optimizes the augmented Lagrangian function $LF_a(\underline{x}^*,\underline{\alpha},\underline{w})$ as well.

Assuming convexity for problem (6.1.1.1.1) and considering equation (6.1.1.1.2) in conjunction with the fact that the penalty terms, i.e. the constraint violations, become zero in the optimal solution point x^*, it follows that the obtained α is identical with the Lagrangian multiplier vector α^*. It can be stated that relative to LF, LF_a with w≥0 has the same Kuhn-Tucker vectors and saddle points. Thus (x^*,α^*) is a saddle point of LF_a if and only if the ordinary Kuhn-Tucker conditions for (6.1.1.1.1) are satisfied.[43]

In the case of nonconvexity of problem (6.1.1.1.1), there may be a duality gap, i.e. the minimal value of the primal problem (6.1.1.1.1) is greater or equal (and not equal, as in the case of convexity) to the maximal solution of its Lagrangian. This is due to the fact that the Lagrangian performs a convexification of its primal problem, also called a Lagrangian relaxation.[44] By changing the Lagrangian function and augmenting it by penalty terms, this duality gap can be eliminated. Rockafellar showed that if $(\underline{x}^*,\underline{\alpha},\underline{w})$ is a saddle point of LF_a for some $\underline{w}≥0$, then $\underline{x}^*$ and $\underline{\alpha}$ satisfy the standard second order necessary conditions for optimality, and $\underline{x}^*$ is (globally) optimal. On the other hand, if $\underline{x}^*$ and $\underline{\alpha}$ satisfy the standard second order necessary conditions and $\underline{x}^*$ is the unique (globally) optimal solution in the strong sense, and the quadratic growth condition is satisfied, then $(\underline{x}^*,\underline{\alpha},\underline{w})$ is a saddle point of LF_a for some $\underline{w}≥0$.[45]

It is convenient to denote the minimizer of the augmented Lagrangian function by $\underline{x}(\alpha)$ and to regard $\underline{w}$ as being fixed during the solution procedure. Now the basic steps of the algorithm can be provided:[46]

Solution procedure with Augmented Lagrangians
Step 1:   determine a sequence $\{\underline{\alpha}^{(k)}\} \rightarrow \underline{\alpha}^*$.

42   Hestenes, M.R.: Multiplier and Gradient Methods, in: JOTA, 4(1969), pp.303-320 und Rockafellar, R.T.: A Dual Approach to Solving Nonlinear Programming Problems by Unconstrained Optimization, in: MP, 5(1973), pp.354-373

43   Rockafellar, R.T.: A Dual Approach to Solving Nonlinear Programming Problems by Unconstrained Optimization, in: MP, 5(1973), pp.361-362

44   Shapiro, J.F.: Mathematical Programming: Structures and Algorithms, New York 1979, p.150

45   Rockafellar, R.T.: Augmented Lagrange Multiplier Functions and Duality in Nonconvex Programming, in: SIAM J. Control, 12(1974)2, p.284

46   see also Fletcher, R.: Methods for Nonlinear Constraints, in: Nonlinear Optimization 1981, Ed.: M.J.D. Powell, London, New York 1982, p.197

Step 2:   for each $\underline{\alpha}^{(k)}$ find a local minimizer $\underline{x}(\underline{\alpha}^{(k)})$ to minimize $LF_\alpha(\underline{x}, \underline{\alpha}^{(k)}, \underline{w})$.
Step 3:   terminate when $c_i(\underline{x}(\underline{\alpha}^{(k)}))$ for $i=\{1,\ldots,m_b\}$ is sufficiently small and $c_i(\underline{x}(\underline{\alpha}^{(k)})) < 0$ $i=\{m_b+1,\ldots,m\}$.

alg. 6.1: Augmented Lagrangian

The problem now consists in finding a sequence of Lagrange multipliers $\underline{\alpha}$, which converge to $\underline{\alpha}^*$. If inequalities in (6.1.1.1.1) are excluded, the adjustment of $\underline{\alpha}$, when $\underline{w}$ is large enough, can be expressed elegantly as a dual problem corresponding to the saddle point problem for LF_α. Let $\Psi(\underline{\alpha})$ be the function:

$$\Psi(\underline{\alpha}) = LF_\alpha(\underline{x}(\underline{\alpha}), \underline{\alpha}, \underline{w})$$

Then the following expression is obtained:[47]

$$\Psi(\underline{\alpha}) \leq LF_\alpha(\underline{x}^*, \underline{\alpha}, \underline{w}) = LF_\alpha(\underline{x}^*, \underline{\alpha}^*, \underline{w}) = \Psi(\underline{\alpha}^*)$$

The second part of the expression depends on the fact that the constraint functions are zero at $\underline{x}^*$. Thus the problem of adjusting $\underline{\alpha}$ is the problem of maximizing the function $\Psi(\underline{\alpha})$.[48]

If Newton's method is applied for this problem, $\underline{\alpha}^{(k+1)}$ can be evaluated in the following form:

$$\underline{\alpha}^{(k+1)} = \underline{\alpha}^{(k)} + (\underline{A}^T \underline{C}^{-1} \underline{A})^{-1} \underline{c} \,\Big|_{x(\underline{\alpha}^{(k)})}$$

Thereby the matrix $\underline{C}$ is the Hessian matrix of the augmented Lagrangian function and matrix $\underline{A}$ the Jacobian matrix of the constraints $\underline{c}$.

6.1.1.2 The Flow-Deviation method

The Flow-Deviation method (FD method) is a general procedure for the solution of non-linear, multicommodity flow problems without additional constraints.[49] In this method the multicommodity flow problem consists of a flow distribution $\underline{x}$ between n nodes, so that each node j obtains a required quantity qc_{ij} of type (i,j) commodity flows from node i. This is done for all commodities so that minimization (maximization) of a well-defined

---

47   Powell, M.J.D.: Algorithms for Nonlinear Constraints that use Lagrangian Functions, in: MP, 14(1978), p.235

48   for a proof that $\underline{\alpha}$ converges to some Kuhn-Tucker vector even if inequalities are considered, see Rockafellar, R.T.: The Multiplier Method of Hestenes and Powell Applied to Convex Programming, in: JOTA, 12(1973)6, pp.558-561

49   Fratta, L., Gerla, M., Kleinrock, L.: The Flow Deviation Method: An Approach to Store-and-Forward Communication Network Design, in: Networks, 3(1973), pp.97-133

performance function is achieved.

Model FD

Decision variables:

$x_{kl}(ij)$: flow from node k to node l between the source node i and sink node j

Parameters:
qc_{ij} : required quantity of type (i,j) commodity at node j from node i

Objective function:
min $P(\underline{x})$ $\qquad$ (6.1.1.2.1)

Constraints:

$$\sum_{k=1}^{n} x_{kl}(ij) - \sum_{m=1}^{n} x_{lm}(ij) = \begin{cases} -qc_{ij} & \text{if } l=i \\ +qc_{ij} & \text{if } l=j \\ 0 & \text{if } l \neq j \text{ or } l \neq i \end{cases} \qquad \forall \ i,j \qquad (6.1.1.2.2)$$

$x_{kl}(ij) \geq 0 \qquad \forall \ k,l,i,j \qquad (6.1.1.2.3)$

In the objective function a performance function $P(\underline{x})$ is minimized (or maximized). The analysis is thereby restricted to multicommodity problems in which the performance depends solely on the global flow $\underline{x}$. In the first constraint set the total flow from source node i to sink node j for commodity (i,j) is fixed to qc_{ij}. This is achieved through equations for the conservation of flows. The last set of constraints ensures non-negative flows.

Procedure for solution

For ease of presentation an equivalent representation of the commodity flows is first introduced. Herein each commodity flow (i,j) is described by a set of {1,..,K} routes $x_k(ij)$, instead of single flows from node to node. Thus each route represents a possible combination of flows from node to node between node i and node j for commodity (i,j).

In the proposed solution procedure the performance function is minimized (maximized) through a allocation in steps of the flow to the shortest (longest) route from node i to node j, i.e to the route that represents the steepest decrease (increase) for $P(\underline{x})$.[50] The shortest (longest) route is evaluated under the metric of the gradient of the performance function. This gradient is defined by:

$$u_k(ij) = \frac{\partial P(\underline{x})}{\partial x_k(ij)}$$

---

50   For the proof of convergence to the optimal solution for a strictly convex performance function see Fratta, L., Gerla, M., Kleinrock, L.: The Flow Deviation Method: An Approach to Store-and-Forward Communication Network Design, in: Networks, 3(1973), pp.127-130

FD algorithm:
Step 0: find a feasible starting vector $\underline{x}^0$ set n=0
Step 1: calculate $\underline{x}^{n+1} = (1 - \tau^*) \cdot \underline{x}^n + \tau^* \cdot \underline{v}$ with $\underline{v}$ having for each commodity (i,j) the required flow qc_{ij} only on its shortest route k': $k' = arg(\min_{k} u_k{}^{(ij)})$ and τ^* being solution of $\min_{\tau} P[(1-\tau) \cdot \underline{x}^n + \tau \cdot \underline{v}]$, $0 \leq \tau \leq 1$
Step 2: if for all $\sum\limits^{K} u_k{}^{(ij)} \cdot (x_k{}^{(ij)n} - v_k{}^{(ij)n}) < \beta$ (β being a given tolerance level, $\beta > 0$): Stop; else set n=n+1 and go to step 1;

alg. 6.2: FD algorithm

6.1.1.3 Greedy algorithms

A greedy algorithm consists of a simple procedure to obtain an optimal or heuristic solution for an optimization problem.

It is called "greedy" because "at every step, the procedure chooses the best morsel it can swallow, without worrying about the future. It never changes its mind: once a candidate is included in the solution, it is there for good; once a candidate is excluded from the solution, it is never reconsidered."[51]

It starts with a set of available candidates C which are used to obtain a solution of the problem, and a set of already chosen candidates S which determine the solution. The latter is an empty set at the beginning of the procedure. Then, at each step a candidate from C is chosen according to a *selection* function and tested for inclusion in the set of chosen candidates S. This is done according to a *feasibility* check, i.e. if the set S with the selected candidate promises to yield a feasible solution, the candidate is removed from C and added to S, if not, it is removed from C but not added to S. Each time the set of chosen candidates is enlarged, it is checked whether the set now constitutes a *solution* of the given problem. If a solution is found or the set C is empty the procedure ends. Otherwise another step is performed by *selecting* a new candidate out of set C.

---

51   Brassard, G., Bratley, P.: Algorithmics Theory and Practice, Englewood Cliffs 1988, p.80

From the above description the major components of a greedy algorithm are given by:

- a set C of available candidates;
- a set S of candidates already chosen;
- a solution function which checks if a particular set of candidates already used provides a solution;
- a feasibility function that checks whether a set of candidates to be used promises a feasible solution;
- a selection function that indicates at any one time which is the most promising of the candidates not yet used, i.e. the candidate of all those available, which changes the objective function in the best way if added to set S of already chosen candidates;
- an objective function, which is the function being optimized.

Greedy algorithm
Step 1:   set C equal to the set of all candidates and S = {}
Step 2:   select the best candidate x according to a selection function   set C = C \ {x}   if S U {x} is feasible then S = S U {x}
Step 3:   if C = {} and S provides no solution then no solution can be generated: stop   if S provides a solution, S is a solution: stop   go to step 2.

alg. 6.3: Greedy algorithm

6.1.2 Queueing networks

These mathematical models analyze the steady-state behavior of a system described as a network of queues. Queueing networks were first developed by Jackson[52], and later extended by Gordon and Newell[53] and Buzen[54]. In a flexible manufacturing system each cell can be considered as a station with a queue and one or more servers, depending on the

52   Jackson, J.: Networks of Waiting Lines, in: OR, 5(1957),pp.518-521
53   Gordon, W.J., Newell, G.F.: Closed Queueing Systems with Exponential Servers, in: OR, 15(1967)2,pp.254-265
54   Buzen, J.: Computational Algorithms for Closed Queueing Networks with Exponential Servers, in: Comm. ACM, 16(1973)9, pp.527-531

number of CNC-machines, load/unload stations etc.. The transportation system works as a central server station which supplies all the other stations with customers (pallets). Queueing networks allow the dynamic system behavior of a flexible manufacturing system to be considered and a reasonable prediction of the performance parameters to be given; e.g. the production rate, the average queue length at each station and the utilization of servers.

Closed queueing networks are characterized by a fixed number of customers (pallets) circulating in the system. They were first applied to flexible manufacturing systems by Solberg.[55] His program, called CAN-Q, uses Buzen's convolution algorithm to analyze product form queueing networks.[56] Whereas CAN-Q only has one pallet type, the program of Solot and Bastos, called MULTIQ, uses an extension which allows several pallet types to be modelled.[57]

Mean value analysis, developed by Reiser and Lavenberg, is an alternative to Buzen's algorithm.[58] It is based on calculations with the first moment (mean value) of statistical distributions. The software package MVAQ by Suri and Hildebrant allows systems with parallel servers and several pallet types to be modelled.[59] Shalev-Oren, Seidmann and Schweitzer developed an extension of the mean value algorithm called PMVA, which incorporates priority queues into the network.[60]

In contrast to closed queueing networks the number of customers (pallets) is not limited in open queueing networks. Parts arrive externally according to a chosen stochastic distribution for the arrival process and leave when completed.[61]

A comparison of different queueing network algorithms with computer simulation based on data from existing flexible manufacturing systems is given by Tempelmeier.[62]

To facilitate the understanding of solution procedures in chapter 8 some insight into the theory of closed queueing networks is given next. It starts with a description of the central server model. Thereafter the convolution model for one customer class is presented in the form in which it was first developed by Gordon and Newell and later applied by Solberg's CAN-Q program for the evaluation of flexible manufacturing systems. Finally a different approach is shown by mean value analysis. Here the version for multiple customer classes, i.e. several pallet types, is discussed.

---

55 Solberg, J.J.: A Mathematical Model of Computerized Manufacturing Systems, in: Proc. 4th Intern. Conf. on Production Research, Tokyo, Japan, Aug.1977, pp.1265-1275

56 Buzen, J.P.: Computational Algorithms for Closed Queueing Networks with Exponential Servers, in: Comm. ACM, 16(1973)9, pp.527-531

57 Solot, P., Bastos, J.M.: MULTIQ: A Queueing Model for FMSs with Several Pallet Types, in: J. Opl. Res. Soc., 39(1988)9, pp.811-821

58 Reiser, M., Lavenberg, S.S.: Mean-Value Analysis of Closed Multichain Queueing Networks, in: J.ACM, 27(1980)2, pp.313-322

59 Suri, R., Hildebrant, R.R.: Modelling Flexible Manufacturing Systems Using Mean Value Analysis, in: J. Manuf. Syst., 3(1984)1,pp.27-38

60 Shalev-Oren, S., Seidmann, A., Schweitzer, P.J.: Analysis of Flexible Manufacturing Systems with Priority Scheduling: PMVA, in: Annals of OR, 3(1985), pp.115-139

61 Buzacott, J.A., Shanthikumar, J.G.: Models for Understanding Flexible Manufacturing Systems, in: AIIE Trans., 12(1980)4, pp.339-350

62 Tempelmeier, H.: Kapazitätsplanung für flexible Fertigungssysteme, in: ZfB, 58(1988)9, pp.963-980

6.1.2.1 The central server model

A queueing network consists of a network of M stations, each station having a queue with s_m servers. Flexible manufacturing systems can be described by a special kind of queueing network called a central server model (fig.7.1). It consists of a central server station m_1 which serves all the other stations $\{m_2,..,m_M\}$ with customers. When a customer finishes service at one of the stations $\{m_2,..,m_M\}$, it is transferred back to the queue of the central server station without any time delay.

In a flexible manufacturing system the transportation system can be interpreted as such a central server station. Each transportation vehicle is a server of the central server station, which supplies the other stations (i.e. the cells, consisting of identical CNC-machines, load/unload stations etc.) with parts. If a part finishes the process at a cell, it must be transported to another cell. Consequently it has to wait for a vehicle to come, to pick it up and to transfer it, i.e. it has to join the queue of the central server station.

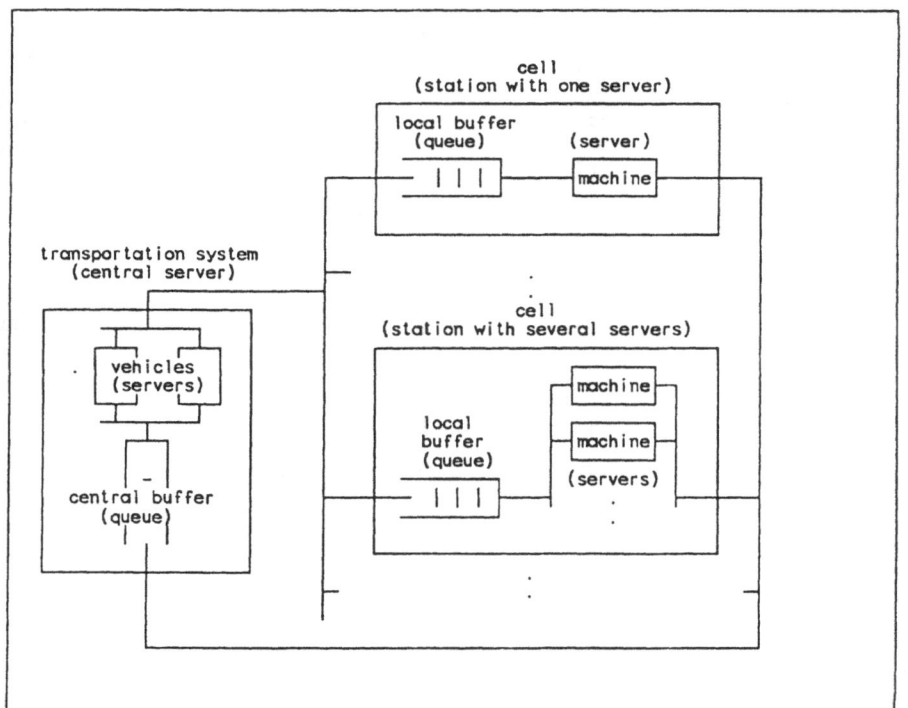

fig. 6.1: Central server model

However, if the real sequences of events in a flexible manufacturing system are compared to those described by the central server model, four differences can be pointed out:

- Unlike in practice the buffers are considered to be unlimited in the central server model.

- The role of the local buffers is different. In the central server model they only serve as input buffers for incoming parts. In practice, however, they can serve as output buffers for the finished parts as well.

- The sequence of transportation processes is different. A part which has completed processing at a station, can in practice be stored in an output buffer at the same station, in the central buffer or in an input buffer of the station where the next operation is performed. In the central server model the finished part is automatically transported with an infinite speed to the queue of the transportation system, i.e. the central buffer. There it remains until a vehicle, i.e. a server of the transportation system, is free to transport it to the next station.

- The role of the central buffers is different. In practice those parts which are neither at one of the cells nor on one of the transportation vehicles must be present in the central buffer. In the central server model they are in the queue of the transportation system. That is why the queue of the transportation system is considered as the central buffer. However, because of the already mentioned difference of transportation sequences, finished parts from one of the cells (stations $\{m_2,..,m_M\}$) need not go to the central buffer. They might be transferred directly to the next station or remain at a local buffer. Thus the performance of the central buffer and of the queue of the central server are different, and, as a result, their marginal queue length distribution deviate from each other.

6.1.2.2 Convolution model for one customer class

The convolution model, as introduced by Gordon and Newell, depends on a few assumptions based on the theory of stochastic processes:[63]

- The system is modeled by a stationary stochastic process;
- Customers (pallets) are stochastically independent;
- Job steps from device to device follow a Markov chain;
- The system is in stochastic equilibrium;
- The service time requirements at each station conform to an exponential distribution;
- The system is ergodic - i.e., long-term time averages converge to the values computed for stochastic equilibrium.

These assumptions are quite restrictive and are violated in practice by flexible manufacturing systems.[64] For example the routes of a pallet through a system are deterministic and do not obey the underlying Poisson process of a Markov chain. Furthermore service times are deterministic and not exponentially distributed. However,

---

63   Denning, P.J., Buzen, J.P.: The Operational Analysis of Queueing Network Models, in: Computing Survey, 10(1978)3, p.226

64   An approach for estimating the properties of queueing systems via product form, but without the restrictive assumptions of stochastic analysis is given by operational analysis. For its assumptions, which pertain to quantities which can be measured from the observation of a system over a finite time horizon, see Denning, P.J., Buzen, J.P.: The Operational Analysis of Queueing Network Models, in: Computing Survey, 10(1978)3, pp.225-261; Dallery, Y., Dubois, D.: L'analyse opérationelle: une approche non

practical experience and comparisons with simulation have shown that closed queueing networks give very good results.[65]

Next the model will be discussed, i.e. its joint probability distribution of system states and the definition of workloads, and then a summary of its properties is given.

6.1.2.2.1 The joint probability distribution of system states

To solve Markovian queueing networks the formulation of a system of balance equations for the joint probability distribution of system states is the traditional approach.

The set of all possible system states can be described by:

$$S(N,M) = \{(n_1, n_2, \ldots, n_M) \mid \text{with} \sum_{m=1}^{M} n_m = N, \; n_m \geq 0, \; m=1, \ldots, M\}$$

Here n_m is the number of pallets (customers) at cell (station) m. Note that each state can be encoded by a binary string of N ones and M-1 zeros:

$$\underbrace{11\ldots10}_{n_1}\underbrace{11\ldots110}_{n_2}\ldots0\underbrace{11\ldots1}_{n_M}$$

The number of strings possible is the number of permutations of N indistinguishable and M-1 indistinguishable objects. In other words, the number of different system states is given by:[66]

$$\frac{(N+M-1)!}{N! \cdot (M-1)!} = \begin{bmatrix} N+M-1 \\ M-1 \end{bmatrix}$$

To obtain the equilibrium probability of a single state $\underline{n} = \{n_1, n_2, \ldots, n_M\}$ Gordon and Newell proved that for closed queueing networks the solution of the balance equations for the system is in the form of a product of simple terms. By normalizing the product terms a proper probability distribution for $\underline{n}$ is derived. It is given by:[67]

$$P(\underline{n}) = \frac{1}{G(M,N)} \prod_{m=1}^{M} f_m(n_m)$$

where $G(M,N)$ is a normalization constant chosen to make all the feasible state probabilities total one:

$$G(M,N) = \sum_{\underline{n} \in S(N,M)} \prod_{m=1}^{M} f_m(n_m)$$

For load-independent stations, i.e. at stations where the service rate μ_m is independent

stochastique des systèmes de files d'attente, in: APII, 20(1986)1, pp.43-86

65 Tempelmeier, H.: Kapazitätsplanung für flexible Fertigungssysteme, in: ZfB, 58(1988)9, pp.963-980

66 Denning, P.J., Buzen, J.P.: The Operational Analysis of Queueing Network Models, in: Computing Survey, 10(1978)3, p.247

67 Gordon, W.J., Newell, G.F.: Closed Queueing Systems with exponential servers, in: OR, 15(1967)2, pp.254-265

of the number of parts n_m (load) at that station, the factor f_m is identical with the relative workloads W_m at station m to the power of the number of pallets n_m at station m.[68]

$$f_m(n_m) = W_m^{n_m} \qquad \text{if station m is load-independent}$$

This is, for example, the case for single-server stations ($s_m = 1$) consisting of a server with a fixed service rate. If, however, a station consists of several servers (multiple-server station) ($s_m > 1$) with a fixed service rate for each server, the station as a whole is now load-dependent due to the fact that the service rate of station m increases with the number of parts n_m, as long as n_m at station m is less than the number of servers s_m. Thus the factor f_m for multiple-server stations is given by:[69]

$$f_m(n_m) = \frac{W_m^{n_m}}{\cdot A_m(n_m)} \qquad \text{if station m is a multiple-server station}$$

with

$$A_m(n_m) = \begin{cases} n_m! & \text{if } n_m \leq s_m \\ s_m! \cdot s_m^{(n_m - s_m)} & \text{if } n_m > s_m \end{cases}$$

To calculate the system parameters with the help of the above probability distribution, the normalization constant $G(M,N)$ must be first evaluated in an efficient way. Buzen developed an algorithm which generates the latter by solving the following formula recursively:[70]

$$C(m,n) = \sum_{k=0}^{n} \frac{W_m^k}{A_m(k)} \cdot C(m-1, n-k)$$

with the starting conditions $\quad C(1,n) = \dfrac{W_1^n}{A_1(n)} \quad \forall \ n \ , \qquad C(m,0) = 1 \quad \forall \ m$

6.1.2.2.2 Workload definitions

As shown in the previous chapter, only the relative values of the workloads are important, and not their absolute value, when ascertaining the equilibrium probability distribution. Thus different ways for their computation exist, of which three different definitions are presented below:

68   Solberg, J.J.: A mathematical model of computerized manufacturing systems, in: Proc. 4th Intern. Conf. on Production Research, Tokyo 1977, p.1269
69   Secco-Suardo, G.: Optimization of closed queueing networks, in: Complex Materials Handling and Assembly Systems Final Report Vol. III No. ESL-FR-834-3, Electr. Syst. Lab. M.I.T., Cambridge MA, July 1978, p.8
70   Buzen, J.P.: Computational Algorithms for Closed Queueing Networks with Exponential Servers, in: Comm. ACM, 16(1973)9, p.529

First Definition:

A very common way to define the relative workload is to use the product of the relative flow e_m with the average processing time t_m at station m:

$$W_m = e_m \cdot t_m$$

Here the relative flow for each station m is derived from flow balance equations using the probability p_{nm} that a pallet after completing service at station n will proceed to station m:[71]

$$e_m = \sum_{n=1}^{M} p_{nm} \cdot e_n \qquad \forall\ m$$

Since this system of M equations is homogeneous, the absolute level of e_m cannot be determined. It can be proven that the throughput at station m is given by:[72]

$$T_m(M,N) = e_m \cdot \frac{C(M,N-1)}{C(M,N)}$$

The throughput of the whole system is then established by dividing the throughput of a station by its average number of visits v_m to complete one part.[73]

$$T(M,N) = \frac{e_m}{v_m} \cdot \frac{C(M,N-1)}{C(M,N)}$$

If the relative flow e_M at the load/unload station is fixed to $e_M = 1$, the throughput of the whole system can be easily derived, provided that the load/unload station is only visited once. The throughput of the load/unload station, which is then equivalent to the throughput of the whole system, is given by the expected production function P(M,N):

$$T_M(M,N) = T(M,N) = P(M,N) = \frac{C(M,N-1)}{C(M,N)}$$

Second definition:

Another formulation of the relative workloads divides the workloads at each station, i.e. the average number of visits v_m multiplied by the average processing time t_m at machine m, by the average workload of each server in the system:

$$W_m = \frac{v_m \cdot t_m}{\sum_{i=1}^{M} v_i \cdot t_i \ / \ \sum_{i=1}^{M} s_i}$$

This definition has two advantages. First the relative workloads are scaled in such a way

71 Baskett, F., Chandy, K.M., Muntz, R.R., Palacios, F.G.: Open, Closed and Mixed Netwoks of Queues with Different Classes of Customers, in: J.ACM, 22(1975)2, p.253

72 see for example Bruell, S.C., Balbo, G.: Computational Algorithms for Closed Queueing Networks, New York and Oxford 1980, pp.49,50

73 Williams, A.C., Bhandiwad, R.A.: A Generation Function Approach to Queueing Network Analysis of Multiprogrammed Computers, in: Networks, 6(1976), p.8

that numerical over and underflows are avoided. The second advantage is that by this formulation the average utilization $U(M,N)$ of the system can be derived very easily, because it is equivalent to the production function $P(M,N)$:[74]

$$U(M,N) = P(M,N) = \frac{G(M,N-1)}{G(M,N)}$$

Then the throughput of the system is given by:

$$T(M,N) = \frac{\sum\limits_{i=1}^{M} s_i}{\sum\limits_{i=1}^{M} v_i \cdot t_i} \cdot \frac{G(M,N-1)}{G(M,N)}$$

Third definition:

In the following a new, different definition of the relative workloads is proposed. The relative workloads W_m are defined as the actual average workloads, i.e. the product of the average number of visits v_{pm} and the average processing time t_{pm} of part type p at machine m weighted by the fraction $frac_p$ part p has of the whole production:

$$W_m = \sum\limits_{p=1}^{P} v_{pm} \cdot t_{pm} \cdot frac_p$$

If the part flow of each part type p is split further according to the fraction of part flow q_{pr} on each alternative route r, one obtains:

$$W_m = \sum\limits_{p=1}^{P} \sum\limits_{r=1}^{R} v_{mpr} \cdot t_{mpr} \cdot q_{pr} \qquad \text{with } \sum\limits_{r=1}^{R} q_{pr} = frac_p$$

This approach can be derived from the first definition if the relative flows e_m are fixed to the average visits v_m at station m. Here the flow balance equations are always satisfied for the central server model. Introducing the definitions for the average processing time t_m and average number of visits v_m to station m:[75]

$$v_m = \sum\limits_{p=1}^{P} v_{mp} \cdot frac_p = \sum\limits_{p=1}^{P} \sum\limits_{r=1}^{R} v_{mpr} \cdot q_{pr}$$

$$t_m = \frac{1}{v_m} \cdot \sum\limits_{p=1}^{P} v_{mp} \cdot t_{mp} \cdot frac_p = \frac{1}{v_m} \cdot \sum\limits_{p=1}^{P} \sum\limits_{r=1}^{R} v_{mpr} \cdot t_{mpr} \cdot q_{pr}$$

the above formula for the relative workloads can be obtained. As in the first approach, the system throughput is given by the production function.

74   Stecke, K.E., Solberg, J.J.: The Optimal Planning of Computerized Manufacturing Systems, School of Industrial Engineering, Purdue University, Report No.20, West Lafayette, Indiana 1981, p.77-78

75   Dallery, Y.: On modelling flexible manufacturing systems using closed queueing networks, in: Large Scale Systems 11(1986),p.113

$$T(M,N) \;=\; P(M,N) \;=\; \frac{G(M,N-1)}{G(M,N)}$$

This formulation has the advantage that it is independent of the number of visits to the load/unload station. Often the latter is a clamping station, which is visited by a part not once but several times for reclamping.

6.1.2.2.3 Properties of the throughput function

When seeking the optimal configuration of a flexible manufacturing system, throughput is the main performance measure of interest.[76] Based on queueing network theory, it was defined in the last chapter as a complex, nonlinear function of several system parameters. In this chapter some recent results concerning the mathematical and qualitative properties of this function will be given.[77] This information provides the necessary information for the search for applicable optimization techniques. For example, in the problem of maximizing the throughput subject to a set of constraints, it is necessary to know if a local maximum is also a global maximum.

Properties in the number of pallets:

Suri proved that a system modelled as a single class closed queueing network with product form is ∞-monotonic, i.e. the throughput of the system is nondecreasing in the number of pallets if the following inequality holds for all stations (cells):[78]

$$t_m(n_m+1) \;\leq\; t_m(n_m) \qquad \forall\; n_m$$

The inequality expresses the fact that the processing time t_m at a cell m is independent (the equality holds) or decreasing with the number of pallets n_m at that cell.[79] For closed queueing network with exponential service times and multi-server stations having unlimited bufferspace, this assumption holds.

Similar results and extensions on this first order property in the number of pallets are given by Whitt[80], Shanthikumar and Yao[81], and Van der Wal[82].

Second-order properties in the number of pallets are examined by Kenvan and von

---

76   see chapter 4.2.2

77   For an overview see also: Shanthikumar, J.G., Yao, D.D.: Second-Order Stochastic Properties in Queueing Systems, in: Proc. of the IEEE, 77(1989)1, pp.162-170

78   Suri, R.: A Concept of Monotonicity and Its Characterization for Closed Queueing Networks, in: OR, 33(1985)3, p.614

79   Even though on the first glance this assumption seems to be reasonable, it does not hold for real systems with large N. There can occur special effects in a flexible manufacturing system due to limited bufferspace, like for example blocking, which leads to an increase in processing time, i.e. an increase in the time a pallet stays at a machine, and consequently causes a decrease in throughput when the number of pallets in the system increases. See Tempelmeier, H., Kuhn, H., Tetzlaff, U.: Performance Evaluation of Flexible Manufacturing Systems with Blocking, in: IJPR, 27(1989)11, pp.1963-1979

80   Whitt, W.: Open and Closed Models for Networks of Queues, in: Bell Lab. Tech. J., 63(1984)9, pp.1911-1979

81   Shanthikumar, J.G., Yao, D.D.: Stochastic Monotonicity of the Queue-Lengths in Closed Queueing Networks, in: Operations Research, 35(1987)4, pp.583-588

82   Van der Wal, J.: Monotonicity of the Throughput of a Closed Exponential Queueing Network in the Number of Jobs, in: OR Spektrum, 11(1989), pp.97-100

Mayrhauser[83]. They proved that throughput is a log convex function of the number of pallets for a single class closed queueing network with cells of instant and single servers. Shanthikumar and Yao provided extensions by considering load-dependent cells[84]. Assuming that the service rate $\mu_m(n_m)$ is a nondecreasing function of n_m at all cells m, they proved that the throughput of a closed queueing network $T(N)$ is a concave (respectively convex, anti-starshaped, starshaped, subadditive or superadditive) function of the job population N if the service rate $\mu_m(n_m)$, as a function of the local queue length n_m, has the same property for each m.

Properties in the number of servers at each station:

Another important property was found by Shanthikumar and Yao. They proved that in a closed queueing network with a set of cells having multiple-servers, i.e. several machines, load/unload stations etc., the throughput is an increasing concave function of the number of servers at each station.[85]

Properties in workload assignments:

So far, much attention has been given to the properties of the throughput function relative to different workload assignments.

First results here were obtained by Price, who proved the convexity of the reciprocal throughput function.[86] Stecke and Solberg proved that the production function $P(M,N,W)$ is strictly quasi-concave but not jointly concave for $N \geq 3$. They further stated the conjecture, based on empirical evidence, that $P(M,N,W)$ is also strongly quasi-concave.[87] By considering a closed queueing network where each station has s servers, Yao and Kim showed that the throughput function is Schur-concave with respect to the loadings.[88] Shanthikumar provided an extension by considering queue dependent server rates $\mu_m \cdot c(n_m)$. Then the throughput is Schur-concave if $c(n_m)$ is increasing and concave.[89]

6.1.2.3 Mean value analysis for multiple customer classes

The convolution method described above evaluates the joint probability distribution. However, for most cases much simpler quantities, such as mean queue size, mean waiting time, utilizations and throughputs, are needed. An alternative approach to the convolution method which avoids the exhaustive calculation of the joint probability distribution and

83   Kenevan, J.R., Mayrhauser von, A.K.: Convexity and Concavity Properties of Analytic Queueing Models for Computer Systems, in: Performance '84, Ed.: E. Gelenbe, Amsterdam 1984, pp.361-375

84   Shanthikumar, J.G., Yao, D.D.: Second-Order Properties of the Throughput of a Closed Queueing Network, in: Math. OR, 13(1988)3, pp.524-534

85   Shanthikumar, J.G., Yao, D.D.: Optimal server allocation in a system of multi-server stations, in: MS, 33(1987)9, pp.1173-1180

86   Price, T.G.: Probability models of multiprogrammed computer systems, Ph.D. Dissertation, Dept. of Electrical Engineering, Stanford University, CA, December 1974

87   Stecke, K.E., Solberg, J.J.: The Optimal Planning of Computerized Manufacturing Systems, School of Industrial Engineering, Purdue University, Report No.20, West Lafayette, Indiana 1981, p.125

88   Yao, D.D., Kim, S.C.: Some Order Relations in Closed Networks of Queues with Multiserver Stations, in: Naval Research Logistics, 34(1987), pp.53-66

89   Shanthikumar, J.G.: Stochastic Majorization of Random Variables with Proportional Equilibrium Rates, in: Adv. Appl. Prob., 19(1987), 854-872

restricts attention to the above-mentioned quantities at the same time is given by mean value analysis. It is based on the relationship between the mean sojourn time and the mean queue size of a system with one customer less. In its general form the mean sojourn time ts_{mc} for pallet type c at station m consists of the mean service time t_m for the pallet itself and the mean service time of the mean queue length seen by the part upon its arrival $n_m(c,\underline{N})$:

$$ts_{mc} = t_m \cdot (1+n_m(c,\underline{N}))$$

Note that in product form queueing networks the service time t_m at each server m is independent of the customer class.

Reiser and Lavenberg showed that for a closed multichain queueing network with product-form solution an arriving pallet of type c observes the equilibrium solution of the queueing network with one less pallet in the arriving pallet's chain :[90]

$$ts_{mc}(\underline{N}) = t_m \cdot (1+\bar{n}_{\underline{m}}(\underline{N-1}_c)) \qquad (6.1.2.3.1)$$

from the arriving pallet of type c observed time average queue length with one pallet of type c, i.e. $(\underline{N-1}_c)$, less in the network

This is due to the fact that the stationary state probabilities at arrival instants are equal to the stationary state probabilities at arbitrary times for the network with one customer less of that type.[91]

Further mean value analysis is augmented by Little's law applied to each customer class and separately solved for each service center:

$$T_c(\underline{N}) = N_c / \sum_{j \in M(c)} ts_{jc}(\underline{N}) \qquad \forall c \qquad (6.1.2.3.2)$$

mean lead time for pallet type c

number of pallets of type c

throughput of type c pallets

$$\bar{n}_{mc}(\underline{N}) = T_c(\underline{N}) \cdot ts_{mc}(\underline{N}) \qquad \forall p, \forall m \in M(c) \qquad (6.1.2.3.3)$$

sojourn time of pallet type c at cell (station) m

throughput of type c pallets

mean number of pallets of type c at cell (station) m

These three formulas (6.1.2.3.1-3) are solved recursively with a incrementation in steps of the pallet number starting with $N_c=0$ $\forall c$. However, the computational requirements increase dramatically when many pallets of many types are considered: if M is the number of stations in the system and N_c the number of pallets, mean value analysis requires $O(M \cdot nN_c)$ calculations. This is the same as the convolution model in its most efficient

---

90   Reiser, M., Lavenberg, S.S.: Mean-Value Analysis of Closed Multichain Queueing Networks, in: JACM, 27(1980)2, pp.316-318

91   Lavenberg and Reiser proved this fact for product form queueing networks with certain types of stations; see Lavenberg, S.S., Reiser, M.: Stationary State Probabilities at Arrival Instants for Closed Queueing Networks with Multiple Types of Customers, in: J. Appl. Prob. 17(1980), p.1056

form.[92] Nevertheless it avoids overflow/underflow problems which may appear in the convolution model (see chapter 6.1.2.2). Furthermore, some interesting heuristic extensions with reduced computational requirements were formulated, of which the one given by Shalev-Oren, Seidmann and Schweitzer is presented next.[93] Its characteristics can be described as follows:

-   Extensive recursion of the exact mean value analysis is avoided by means of an iterative procedure. Thus the difference between arrival-average and time-average queue lengths having N pallets in the system is approximated by means of an heuristic correction factor. As already pointed out above, under certain conditions this difference is equivalent to the time average difference between the mean queue sizes of the system with population N and $N\text{-}1_c$. Instead of recursively solving the above exact mean-value formulation starting with a pallet distribution with $N_c = 0$ $\forall$c, the actual pallet distribution is taken and adjusted by this heuristic correction factor. For the latter a couple of different suggestions exist in literature.[94] The formulas thus obtained are then solved iteratively until a certain convergence parameter is met.

-   It uses an heuristic extension for class-dependent service times. Product form queueing networks are based on the assumption that service times are independent of the customer class. Numerical results show, however, that the modification to class dependent service times is practicable.[95]

-   The algorithm allows stations with several parallel servers to be analyzed, i.e. cells with several identical machines.

-   The mean sojourn time is split into three quantities:

mean service time of the arriving pallet,

mean service time of the mean backlog of pallets waiting upon arrival,

mean residual time of the pallet under process upon arrival. (In the multiple-server case the smallest mean residual time is considered.)

This allows Shalev-Oren et al. to introduce different queueing disciplines, i.e. AS (ample-server), FCFS (first-come-first-serve) or HOL (head-of-line) non-preemptive policy.

The above exact mean value analysis algorithm is then transformed into two basic formulas, which are solved iteratively. In the first one the throughput T_{mc} for each pallet type c at each cell m is calculated by:

92   Reiser, M., Lavenberg, S.S.: Mean-Value Analysis of Closed Multichain Queueing Networks, in: JACM, 27(1980)2, p.319

93   Shalev-Oren, S., Seidmann, A., Schweitzer, P.J.: Analysis of Flexible Manufacturing Systems with Priority Scheduling: PMVA, in: Annals of OR, 3(1985), pp.115-139; see also Schweitzer, P.J.: Approximate Analysis of Multiclass Closed Networks of Queues, in: Proc. Intern. Conf. on Stochastic Control and Optimization, Amsterdam 1979, pp.25-29;

94   an overview is given in: Schweitzer, P.J., Seidmann, A., Shalev-Oren, S.: The Corrections Terms in Approximate Mean Value Analysis, in: Operation Research Letters, 4(1986)5, pp.197-200;

95   Hildebrant, R.R.: Scheduling Flexible Machining Systems using Mean Value Analysis, in: Proc. IEEE Conf. on Decision and Control, Albuquerque, New Mexico 1980, p.706

$$T_{mc}(\underline{N}) = N_c / \sum_{j=1}^{M} \frac{v_{jc}}{v_{mc}} \cdot (tw_{jc} + t_{jc}) \qquad \forall \ c,m$$

$$\left\lfloor \quad \left\lfloor \quad \underbrace{}_{\text{mean lead time for pallet type c}} \right. \right.$$

$$\left. \quad \right\rfloor \text{ number of pallets of type c}$$

$$\left\lfloor \text{ throughput of type c pallets at station (cell) m} \right.$$

The second formula specifies the mean time spent in queue tw_{mc} and has for the FCFS-discipline the following form:

$$tw_{mc} = tw_{Omc} + (1/s_m) \ [\ \sum_{j=1}^{C} (T_{mj} \cdot tw_{mj} \cdot t_{mj}) - T_{mc} \cdot tw_{mc} \cdot t_{mc}/N_c\ ] \qquad \forall \ c,m$$

$$\left\lfloor \quad \left\lfloor \quad \underbrace{}_{\text{mean service time of the mean backlog of pallets waiting upon arrival}} \right. \right.$$

$$\left. \quad \right\rfloor \text{ mean residual time for the pallet under process freed next upon arrival}$$

$$\left\lfloor \text{ mean time of a pallet of type c spent in queue of station (cell) m} \right.$$

Hereby tw_{mc} consists of the mean service time of the mean backlog of pallets waiting upon arrival and the mean residual time tw_{Ocm} for the pallets under process freed next upon arrival.

6.1.3 Computer simulation

There are a variety of definitions for simulation.[96] One example is that given by Pritsker for computer simulation: "Simulation is the establishment of a mathematical-logical model of a system and the experimental manipulation of it on a digital computer."[97]

A simulation study can be divided into ten steps.[98]

1. Formulate the problem and plan the study.

2. Collect data and define a model. Data is collected and then used to estimate input parameters and to obtain probability distributions for the random variables used in the model. The construction of a mathematical and logical model of a real-world system for a given objective is still much more an art than a science.

For the building of models for flexible manufacturing systems Carrie gives an extensive overview.[99] Tempelmeier and Endesfelder developed a module processor which is a tremendous aid to model building and programming. A library of predefined modules in the simulation language SIMAN for flexible manufacturing systems is available. Its flexibility allows further extensions of the library with new modules developed by

96   for an overview of different definitions see: Weber, K., Trzebiner, R., Tempelmeier, H.: Simulation with GPSS - Lehr- und Handbuch mit wirtschaftswissenschaftlichen Anwendungsbeispielen, Bern, Stuttgart 1983, pp.31-32

97   Pritsker, A.A.: The GASP IV Simulation Language, New York 1974, p.1

98   See Law, M.A., Kelton, W.D.: Simulation Modeling and Analysis, New York 1982, pp.43-46; other descriptions can be found in: Weber, K., Trzebiner, R., Tempelmeier, H.: Simulation with GPSS - Lehr- und Handbuch mit wirtschaftswissenschaftlichen Anwendungsbeispielen, Bern, Stuttgart 1983, pp.34-52; Banks, J., Carson, J.S.: Discrete-Event System Simulation, Englewood Cliffs 1984, pp.11-15;

99   Carrie, A.: Simulation of Manufacturing Systems, Chichester 1988, pp.183-396

users.[100]

3. Validation of the model.

4. Construction of the computer program and verification.

5. Make pilot runs.

6. Validation with the help of the output gained from pilot runs.

7. Design experiments. It must be decided what system design to simulate if there are more alternatives than one can reasonably simulate.

8. Carry out production runs.

9. Analyze output data.

10. Document and implement results. Because simulation models are often used for more than one application, it is important to document the assumptions of the model and the program as well.

Thus simulation permits the evaluation of desired system parameters by modelling and experimental manipulation. It is more flexible then analytical tools, which are a priori based on certain model assumptions. For some design problems, simulation is the only tool applicable, since there are no adequate analytical tools available. However, modelling, programming and running simulations are time consuming and require expertise.

6.1.4 Perturbation analysis

Perturbation analysis is a technique for the computation of gradients of performance measures for a discrete event dynamic system.[101] This system can be a "real world" flexible manufacturing system[102] or a simulated flexible manufacturing system[103].

The basic idea is to observe a given sample path, called the nominal path, and to consider the following question: if the time of occurrence of a specific event in the nominal path had been perturbed, what would have been the effect on the various performance measures at the end of the observation period? To answer this question each perturbation generated by so-called perturbation generation rules is propagated along the nominal path according to propagating rules, and the various perturbations are added together using superposition rules to determine the effect on the performance measures. With a little more computational effort these additional calculations can be implemented on a single simulation run or experiment.

---

100 Tempelmeier, H., Endesfelder, T.: Der SIMAN MODUL PROZESSOR - ein flexibles Softwaretool zur Erzeugung von SIMAN-Simulationsmodellen, in: Angewandte Informatik, 29(1987)2, pp.104-110; Endesfelder, T., Tempelmeier, H.: The SIMAN Module Processor - A Flexible Software Tool for the Generation of SIMAN Simulation Models, in: Simulation in CIM and Artificial Intelligence Techniques, Ed.: Retti, J., Wichmann, K.E., Proc. European Simulation Multiconference, July 1987, Vienna

101 Ho, Y.C.: Perturbation Analysis explained, in: Working Paper Division of Applied Sciences, Harvard University, 1987, p.1

102 Suri, R., Dille, J.W.: A Technique for on-line Sensitivity Analysis of Flexible Manufacturing Systems, in: Annals of OR, 3(1985), pp.381-391

103 Suri, R.: Implementation of Sensitivity Calculations on a Monte Carlo Experiment, in: JOTA, 40(1983)4, pp.625-630; Suri, R., Cao, X.: Optimization of flexible manufacturing systems using new techniques in discrete event systems, in: Proc. 20th Allerton Conf. Communic. Control and Computing, Monticello, Illinois 1982, pp.434-443

6.1.5 Petri nets

The Petri net is a tool which allows the analysis of parallel, concurrent processes. The "normal" Petri net, as first introduced by Petri, consists of two types of nodes, places and transitions joined by arcs.[104] The set of places represents the states of the system, while the set of transitions represents the events. The basic tools to analyze certain properties, such as boundedness and liveness of Petri nets, are the incidence matrix and the reachability graph.[105] Unfortunately the size of a Petri net can become very large if a complex system such as a flexible manufacturing system is analyzed. Also slight functional changes of the system can substantially change the structure of the Petri net.

Colored Petri nets allow a more compact representation. However, this complicates further analysis and even some of the techniques applicable for Petri nets are not yet fully developed for colored Petri nets.

If system characteristics like utilization and throughput have to be evaluated, timed Petri nets or stochastic Petri nets are applicable. These results can be obtained with the help of simulation of timed Petri nets.[106] If the timed Petri net is decision-free, efficient algorithms based on graph-theoretic concepts exist.[107] Alternatively, if stochastic data is used, a stochastic Petri net allows the evaluation of the above mentioned performance measures with the help of an isomorphic Markov chain.[108]

An extensive bibliography of Petri net literature is given in Drees et al.[109]

6.2 Domain specific tools

6.2.1 Group technology methods

Group technology comprises a number of methods for analyzing and arranging the part spectrum and the relevant manufacturing processes according to similarities (e.g. geometry, size, volume, type and sequence of operation, etc.). The goal is to establish a number of production cells, each of which is capable of producing a group or family of components.[110] If group technology is applied to the design of flexible manufacturing systems, a group technology cell can be interpreted as a system of this type.

104   Petri, C.A.: Kommunikation mit Automaten, Schriften der Rheinischen Westfälischen Instituts für instrumentelle Mathematik an der Universität Bonn, Bonn 1962

105   Martinez, J., Alla, H., Silva, M.: Petri Nets for the specification of FMSs, in: Modelling and Design of Flexible Manufacturing Systems, edited by A. Kusiak, Amsterdam 1986, pp.389-406; Narahari, Y., Viswanadham, N.: A Petri Net Approach to the Modelling and Analysis of Flexible Manufacturing Systems, in: Annals of OR, 3(1985), pp.449-472

106   Ravichandran, R: Decision Support in FMS Using Timed Petri Nets, in: J. Manuf. Syst., 5(1986)2, pp.89-101

107   Cohen, G., Moller, P., Quadrat, J.P., Viot, M.: Linear System Theory for Discrete Event Systems, in: Proc. 23nd IEEE Conf. on Decision and Control, Las Vegas 1984, pp.539-544

108   Molloy, M.K.: On the Integration of Delay and Throughput Measures in Distributed Processing Models, Ph.D. Thesis, University of California, Los Angeles 1981,p.34

109   Drees, S., Gomm, D., Plünnecke, H., Reisig, W., Walter, R.: Bibliography of Petri Nets, in: Lecture Notes on Computer Sciences, No.266, Heidelberg, 1987, pp.309-451

110   Chakravarty, A.K., Shtub, A.: An integrated layout for group technology with in-process inventory costs, in: IJPR, 22(1984)3, p.431

Ballakur classifies the techniques in group technology into three classes:[111]

- Form part families and then group machines into cells (part family grouping).

  Here classification and coding schemes, such as the one of Opitz can be subsumed.[112] An overview on coding systems can be found in Hyer and Wemmerlöv.[113] Further cluster analysis can be applied to find part families.[114]

- Form machine cells based on similarity in part routings and then allocate parts to cells (machine grouping).

  Again cluster analysis, such as the single linkage cluster analysis, can be applied.[115] Another approach for machine grouping are graph-theoretical methods like the one of Rajagopalan and Betra.[116] Olivia-Lopez and Purcheck perform machine grouping based on combinatorial analysis.[117]

- Form part families and machine cells simultaneously (machine-part grouping).

  Manual techniques, such as production flow analysis of Burbidge[118] and component flow analysis of El-Essawy[119], perform such a simultaneous grouping of parts and machines. Furthermore a number of algorithmic techniques exists which are based on matrix reordering of a machine-part matrix, e.g. the rank order clustering algorithm of King[120] or the bond energy algorithm of McCormick et al..[121] Some newer approaches also include cost considerations. Here for example those of Seifoddini[122], of Askin and Subramanian[123] and of Mutti and Semeraro[124] can be listed.

For a more in depth review of group technology methods refer to theses by Ballakur[125] or Seifoddini[126] or papers by Huang and Houck[127] or Mosier and Taube[128].

111   Ballakur, A.: An Investigation of Part Family/Machine Group Formation for Designing Cellular Manufacturing Systems, Ph.D. Thesis University Wisconsin, Madison 1985, pp.10-38
112   Opitz, H.: Verschlüsselungsrichtlinien und Definitionen zum werkstückbeschreibenden Klassifizierungssystem, Essen 1966
113   Hyer, N.L., Wemmerlöv, U.: Group Technology oriented Coding Systems: Structures, Applications, and Implementation, in: Production and Inventory Management, (1985)2, pp.55-78
114   e.g. Carrie, A.S.: Numerical Taxonomy Applied to Group Technology and Plant Layout, in: IJPR, 11(1973)4, pp.399-416
115   McAuley, J.: Machine Grouping for Efficient Production, in: The Production Engineer, 51(1972)2,pp.53
116   Rajagopalan, R., Betra, J.L.: Design of Cellular Production Systems: A Graph-Theoretic Approach, in: IJPR, 13(1975)6, pp.567-579
117   Olivia-Lopez, E., Purcheck, G.F.: Load Balancing for Group Technology Planning and Control, in: Int. Journal of Machine Tool Design and Research, 19(1979)4, pp.259-274
118   Burbidge, J.L.: Production Flow Analysis, in: The Production Engineer, 42(1963)12, pp.742-752
119   El-Essawy, I.F.K.: The Development of Component Flow Analysis as a Production Systems' Design for Multi-Product Engineering Companies, Ph.D. Thesis, UMIST, U.K. 1971
120   King, J.R.: Machine-Component Grouping in Production Flow Analysis: An Approach Using Rank Order Clustering Algorithm, in: IJPR, 18(1980)2, pp.213-232
121   McCormick, W.T., Schweitzer, P.J., White, T.E.: Problem Decomposition and Data Recognition by a Clustering Technique, in: OR, 20(1972),pp.993-1009.
122   Seifoddini, H.: Cost based machine-component grouping model: in Group Technology, Ph.D. Thesis Oklahoma State University, Stillwater 1984
123   Askin, R.G., Subramanian, S.P.: A cost-based heuristic for group technology configuration, in: IJPR, 25(1987)1, pp.101-113
124   Mutti, R., Semeraro, Q.: A heuristic method to group pieces for economical production, Working Paper WP-1987-001-QS, Politecnico di Milano, dipartimento di meccanica, Milano 1987
125   Ballakur, A.: An Investigation of Part Family/Machine Group Formation for Designing Cellular Manufacturing Systems, Ph.D. Thesis University Wisconsin, Madison 1985, pp.10-38
126   Seifoddini, H.: Cost based machine-component grouping model: in Group Technology, Ph.D. Thesis Oklahoma State University, Stillwater 1984, pp.22-39

6.2.2 Knowledge-based systems

A knowledge-based system denotes a computer program that can store knowledge of a particular domain and use that knowledge to solve problems from this domain in an intelligent way.[129] Knowledge-based systems are also called expert systems if they refer to problem solving at that level of performance and in those areas that are usually achieved by human experts.

According to Kriz a knowledge based system contains at least the following two components:[130]

- knowledge base. Here the domain-specific knowledge is stored in general in the form of facts or rules.

- inference engine. It operates on the knowledge-base, performs logical inferences and deduces new knowledge by applying rules to facts until the posed problem is solved.

Besides these basic components it should provide the following features:

- explanatory capabilities of its behavior on request by the user,

- a user-friendly dialogue,

- an application-oriented knowledge representation language,

- the possibility to easily modify and extend the knowledge-base during the lifetime of the system.

A knowledge-based system with the above listed features, but an empty knowledge base is called an expert system shell. By supplying it with the particular knowledge, a specific knowledge-based system can be built. In Assad and Golden an overview of shells applicable for microcomputers is given.[131]

An overview of already existing systems in the development, implementation and operation of automated manufacturing systems is given by Kusiak and Heragu.[132] Mellichamp and Wahab developed an expert system which modifies the design of a flexible manufacturing system and uses simulation to evaluate the design.[133]

7. Modelling the basic planning stage

The remainder of this thesis will concentrate on the basic planning stage. In this

127 Huang, P.Y., Houck, B.L.W.: Cellular Manufacturing: An Overview and Bibliography, in: Production and Inventory Management, (1985)4, pp.83-93

128 Mosier, C., Taube, L.: The Facets of Group Technology and Their Impacts on Implementation - A State-of-the-Art Survey, in: OMEGA, 13(1985)5, pp.381-391

129 Hayes-Roth, F., Waterman, D.A., Lenat, D.B.: Building Expert Systems, Reading Mass. 1983, pp.3,4,31-57

130 Kriz, J.: Knowledge-based systems in industry: Introduction, in: Knowledge-based systems in industry, editor J. Kriz, Chichester 1987, p.11

131 Assad, A.A., Golden, B.L.: Expert Systems, Microcomputers, and Operations Research, in: Comp. & Ops. Res., 13(1986)2/3, pp.301-321

132 Kusiak, A., Heragu, S.S.: Expert Systems and Optimization in Automated Manufacturing Systems, Working Paper No. 07/87, University of Manitoba, Departement of Mechanical and Industrial Engineering, May 1987

133 Mellichamp, J.M., Wahab, A.F.A.: An expert system for FMS design, in: Simulation,48(1987)5,pp.201-208

chapter an insight is given into important aspects of modelling the basic planning stage.

First the major role of mathematical programming during this stage is explained.

Afterwards criteria for the classification of models for the basic planning stage are given (see fig. 7.1).

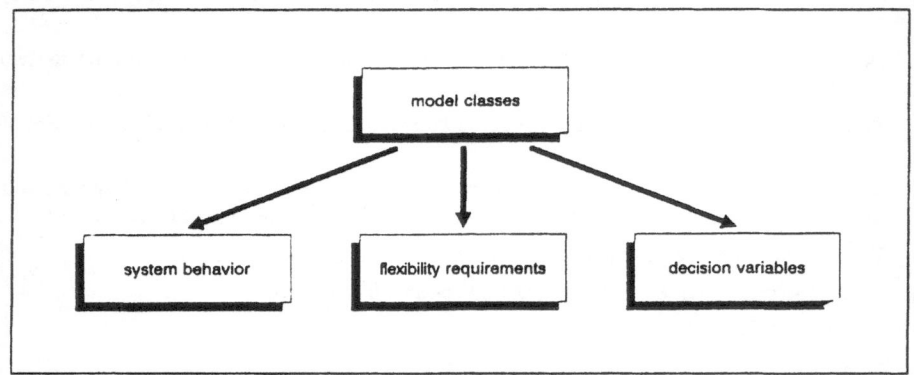

fig. 7.1: Classification criteria for models of the basic planning stage

Based on the fact that system behavior can be modelled in a dynamic or static approach, two model classes are distinguished and some interesting relationships between these classes are shown. Also flexibility requirements necessitate certain model characteristics which allow further model classes to be distinguished. A third classification of model classes can be derived according to the decision variables used.

Finally this chapter ends with the description of a cost model for analyzing the relevant costs to be considered during the basic planning stage.

7.1 Mathematical programming as the dominating tool

When the tools for the design process are analyzed with respect to their applicability for the basic planning stage, it first becomes apparent that some techniques, i.e. queueing networks, simulation and Petri nets only provide results of an evaluative character. That is, they do not find, but only evaluate a solution and its characteristics. However, they can be used in combination with other, generative techniques to improve a solution. Mathematical programming, knowledge-based systems and methods from group technology can be identified as techniques with capabilities to find solutions for the basic planning stage. Perturbation analysis can be considered as evaluative with semi generative properties.[134]

Below mathematical programming is referred to as the dominant tool for solving problems at the basic planning stage. This is because it allows quantitative optimization models to be developed whose goal it is to find an optimal solution according to a given objective function. Unfortunately the models so far published in literature have limited

---

134 Suri, R.: An overview of evaluative models for flexible manufacturing systems, in: Annals of OR, 3(1985),p.16

capabilities. Thus some new models to extend these capabilities will be presented later.

The alternatives, i.e. knowledge-based systems and methods from group technology, have some limitatiors, which, in the opinion of the author, allow them to play only a supplementary role during the design process.

Knowledge-based systems are only capable of producing a feasible solution through an inference engine, which combines a given set of rules according to a given problem specification. Based on this solution strategy, mainly qualitative knowledge is applicable. However, the objective and the character of the basic planning stage emphasizes quantitative techniques leading to an optimal or at least a satisfactory solution.

In group technology those techniques which simultaneously form part families and machine cells (machine-part grouping) or only part families (part family grouping) seem to be useful for the basic planning stage.

The methods for machine-part grouping can be seen as approaches to generate a solution for the basic planning stage. However, machine-part grouping must be seen in its original context, and the objective for which they were first developed should not be forgotten. Originally their goal was to redesign an existing job shop with the aim of combining the job shop with the advantages of a flow shop. Therefore cost considerations were neglected in the earliest approaches, such as production flow analysis, component flow analysis, or matrix reordering techniques like the rank order clustering algorithm. In some newer techniques such as those developed by Seifoddini, Askin and Subramanian, and Mutti and Semeraro cost considerations are incorporated.[135] The latter in particular explicitly considers a new production system and not the reorganization of an old one.[136] However, there are still three significant arguments against the use of machine-part grouping for the basic planning stage:

- Firstly, only cells (flexible manufacturing systems) and no alternative production systems, for example a flow shop or a job shop, are generated.

- Secondly, alternative routes of a part type, caused by alternative process plans and/or by the fact that one single operation of a part type can be performed on different alternative machines are not considered. Due to the flexibility of modern CNC-machines this is however an important factor, which allows better utilization of resources and hence less equipment.

- The last argument is that changes in production requirements and flexibility considerations are in general neglected by the existing methods.

Compared to machine-part grouping, part-family grouping seems to be more applicable during the basic planning stage. There it can be used as a complementary tool to mathematical programming. First part-family grouping groups the parts into part families or into part groups for productions systems. Afterwards decision models based on mathematical programming generate a system configuration for each family, or a selected

135   Seifoddini, H.: Cost based machine-component grouping model: in Group Technology, Ph.D. Thesis Oklahoma State University, Stillwater 1984; Askin, R.G., Subramanian, S.P.: A cost-based heuristic for group technology configuration, in: IJPR, 25(1987)1, pp.101-113; Mutti, R., Semeraro, Q.: A heuristic method to group pieces for economical production, Working Paper WP-1987-001-QS, Politecnico di Milano, dipartimento di meccanica, Milano 1987

136   Mutti, R., Semeraro, Q.: A heuristic method to group pieces for economical production, Working Paper WP-1987-001-QS, Politecnico di Milano, dipartimento di meccanica, Milano 1987, p.5

number of part groups.

7.2 The relationship between dynamic and static system behavior modelling

The objective of this chapter is to provide some insight into the relationship between dynamic and static system behavior modelling.

Dynamic modelling is achieved primarily by including the queueing processes in front of the cells into the modelling consideration. Additionally aspects of the part flow between the cells are usually incorporated (see fig. 7.2). As already mentioned in chapter 6.1.2 dynamic system behavior can be quite accurately captured through queueing network theory. It allows the derivation of the system's throughput and other performance measures.

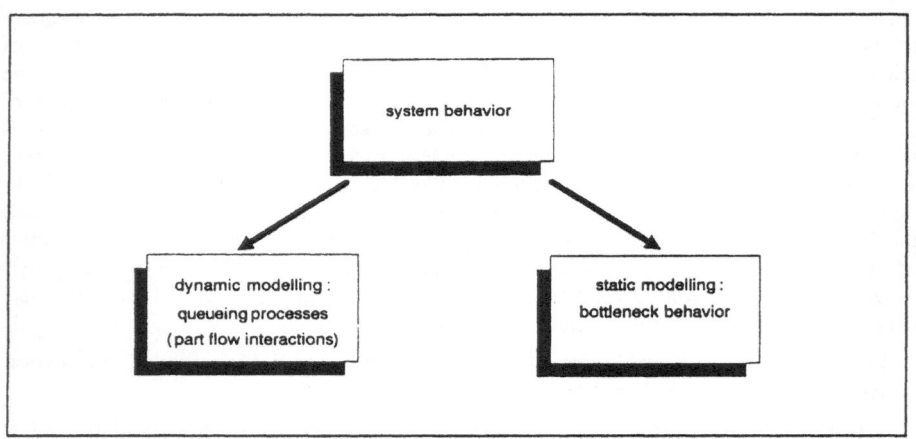

fig. 7.2: Model classes according to system behavior modelling

Static modelling, which neglects queueing processes and important aspects of the part flow, is achieved by comparing the available capacity of a station (cell or transportation system) to its workload. The throughput of a station (cell or transportation system) is then determined by the workload per server (CNC-machine, load/unload station, vehicle etc.), i.e. the workload W_m at station m divided by the number of servers s_m or equivalently the number of visits v_m divided by the service rate μ_m and the number of servers s_m. Moreover, the system's throughput can be determined by the bottleneck station b, i.e. the station (cell or transportation system) with the highest workload per server.

$$\frac{W_b}{s_b} = \frac{v_b}{\mu_b \cdot s_b} = \max_m \frac{v_m}{\mu_m \cdot s_m}$$

Hence the bottleneck station is fully utilized. The fact that the system's maximal throughput is given by the capacity of the bottleneck station, allows simple linear programming models to be constructed. These are much easier to solve compared with mathematical programming models, which comprise nonlinear queueing network formulas. However, the question arises to what extent the throughput evaluation by static modelling differs from dynamic modelling of system behavior. If dynamic modelling is achieved by a standard closed queueing network of the Jackson type[137], it can be observed that the system's throughput reaches a saturation point if the number of customers (pallets) increases and becomes very large. This point can be determined as follows. According to Shanthikumar and Yao it can be proved, that the throughput in a standard closed queueing network is an increasing function with an increasing number of pallets N, i.e. $T(N_1) \leq T(N_2) \leq T(N_3)....$ with $N_1 < N_2 < N_3....$[138] A system with this property is defined as ∞-monotone1. Due to flow balance equations (see chapter 6.1.2.2.1) at each station, the throughput of any station cannot exceed its (maximal) service rate $\mu_m \cdot s_m$. Hence if the number of pallets in the system is unlimited, the throughput of the system is determined by the bottleneck station, the station (cell or transportation system) with the highest workload per server. Thus the saturation point of the closed queueing network is identical to the throughput of the bottleneck station:[139]

$$\lim_{N \to \infty} T = \frac{e_b}{v_b} \cdot \frac{G(M,N-1)}{G(M,N)} = \frac{\mu_b \cdot s_b}{v_b} = \frac{s_b}{W_b}$$

and it can be concluded that, when a very large number of pallets is in the system, dynamic throughput behavior is almost identical to static throughput behavior modelling.

This fact leads us to another interesting relationship between static and dynamic modelling when server costs are considered. The following theorem can be stated:

Theorem 1: If increasing cost functions for servers (CNC-machines, load/unload stations, transport vehicles etc.) $\underline{s}$ and pallets N are assumed, i.e. $g(\underline{s})$ and $g(\underline{s},N)$ are increasing, static modelling with cost minimization of a system for a given throughput R_{min}, i.e.

$F_S = \min \quad g(\underline{s})$

$\quad s.t. \quad R_{min} \cdot W_m \leq s_m \quad \forall m$

yields a lower bound solution for the server costs of the equivalent dynamic model evaluating the throughput $T(\underline{W},\underline{s},N)$ by standard closed queueing network theory, i.e. $F_S \leq F_D$ with the dynamic model given by

$F_D = \min \quad g(\underline{s},N)$

$\quad s.t. \quad T(\underline{W},\underline{s},N) \geq R_{min}$

Proof: Based on the above-stated equivalence between the throughput obtained through static modelling, i.e. the throughput of the bottleneck station (cell or transportation system), and the saturation point of closed queueing networks, it can be concluded, that the solution for minimal server costs found by static modelling is equivalent to the solution for minimal server costs obtained by closed queueing network theory with an unlimited number of pallets in the system. If now those costs which depend on the number of pallets, i.e. pallet costs, fixture costs and inprocess inventory costs, are not neglected, the number of pallets becomes limited. Hence, due to ∞-monotonicity, the throughput of the system is

---

137   see chapter 6.1.2.2

138   Shanthikumar, J.G., Yao, D.D.: Stochastic Monotonicity of the Queue-Lengths in Closed Queueing Networks, in: Operations Research, 35(1987)4, p.585

139   Secco-Suardo, G.: Optimization of closed queueing networks, in: Complex Materials Handling and Assembly Systems Final Report Vol. III No. ESL-FR-834-3, Electr. Syst. Lab. M.I.T., Cambridge MA, July 1978, pp.19-22

decreasing. However, as proved by Shanthikumar and Yao, it must borne in mind that in a closed queueing network with a set of multi-server stations, the throughput is an increasing concave function of the number of servers at each station.[140] Thus the server configuration $\underline{s}$ remains constant or must be increased by adding new servers to fulfill production requirements, if the number of pallets is limited.

7.3 Model classes according to flexibility requirements

A further classification of models can be derived from flexibility requirements. In chapter 2.3 it was shown that production demand over a longer time horizon (i.e. the part family), and the production demand over an infinitesimally short time horizon (i.e. the part mix), determine the required flexibilities. As is shown in the following, the characteristics of the part mix and the part family have considerable influence on modelling requirements.

According to Kleinrock two different classes of flows in systems can be distinguished, i.e. steady and unsteady flows.[141] The first class consists of those systems in which the flow proceeds in a predictable fashion, i.e. the quantity of flow is exactly known and is constant over the interval of interest. Thus the time when the flow appears at the route, and how much of a demand that flow places upon the route is known and constant. In contrast the second class consists of stochastic flows. Here the times at which demands for service arrive (use of a route) are uncertain or unpredictable, and also the size of the demands themselves that are placed upon the route are unpredictable.

Based on this distinction it can be stated that for an irregularly changing part mix with many part groups, as observed in flexible machining systems, a stochastic model is appropriate. On the other hand, flexible transfer lines or multi-lines show a regular part mix with only one or a few part groups and with a predictable flow of parts with no or limited interactions between the flows of different part groups. For these kinds of systems with a steady flow, deterministic models can be applied.

The characteristics of the part family allows further distinction between production requirements which remain constant over a long time horizon, and those which show a changing level. If for both extreme types of flexible manufacturing systems, i.e. a flexible machining system and a flexible transfer line, changing production levels in the part family are observed over a longer time horizon, multi-period models are appropriate. They allow the partition of production requirements in such a way that for each single period a constant part family can approximately be assumed. On the other hand, a constant level of production requirements only requires a single-period model (see fig.7.3).

---

140   Shanthikumar, J.G., Yao, D.D.: Optimal server allocation in a system of multi-server stations, in: MS, 33(1987)9, pp.1173-1180
141   Kleinrock, L.: Queueing Systems, Volume I: Theory, New York, Toronto 1975, pp.4-6

flexibility requirements → part mix ↓ part family	irregular with many part groups (flexible machining system)	regular with one or few part groups (flexible transfer line or multi-line)
constant	stochastic single-period models	deterministic single-period models
changing level	stochastic multiperiod models	deterministic multiperiod models

fig. 7.3: Modelling according to flexibility requirements

Based on these observations flexible machining systems are best modelled provided dynamic system behavior modelling based on stochastic queueing theory is applied. For modelling the steady flow in flexible transfer lines and multi-lines, static system behavior modelling, which neglects queueing processes and interactions between the different part flows, might suffice. This guideline can be supported by the observations of Bondi and Whitt. They noticed that the more spread out or dispersed the service time and arrival processes are in some queueing networks, the more the congestions and thus the mean queue lengths increase.[142]

Furthermore, if the part family of a flexible manufacturing system shows changing levels in production requirements, it is necessary to use multiperiod models.

7.4 Model classes according to decision variables

Another distinction between models is due to the capacity for performing different design decisions based on the use of different decision variables incorporated in the models. The decisions at the basic planning stage (see chapter 5.1) are reduced to three basic categories of decision variables:

- variables for part routing/operation assignment. Here the part routing or the assignment of operations is determined. Based on a given required production rate for each part type, it is necessary to decide how much of each part type is produced on each of its alternative routes, or which operation to perform on which resource.

- variables for server decisions. These decision variables describe how many CNC-machines, load/unload stations, vehicles for the transportation system etc. are to be implemented in the system.

- variables for pallet decisions. In some models with a limited number of pallets, decisions about the number of pallets are possible. Here a further distinction can be made, depending on whether only one pallet type is used or if several different pallet types are considered.

Note that some of the basic categories of decision variables can be aggregated, e.g. server and pallet decisions, to a system decision variable. Further aggregations in each category are possible, e.g. operations of a part type to operations of a part family.

---

142 Bondi, A.B., Whitt, W.: The Influence of Service-Time Variability in a Closed Queueing Network of Queues, in: Performance Evaluation, 6(1986), pp.219-234

Based on the capability of the models and their decision variables the following classification of optimization models is derived.

Routing optimization: Here the routing of parts in a given flexible manufacturing system is optimized. Therefore in this category only decision variables concerning the part routing/operation assignment are found. As the system is already configured, models of this type are not directly applicable for design optimization. However, routing optimization is integrated in quite a few complexer models.

Capacity optimization: The necessary types of equipment must have already been selected in order to optimize capacity. Furthermore the part routing through the system is given. Now the optimal quantity of each type of equipment must be found necessary to satisfy a certain objective and obeying certain restrictions at the same time. As a result variables for CNC-machines / load-unload stations / transportation system decisions and/or variables for pallet decisions are used.

Equipment optimization: The models in this class are not only able to determine the optimal amount of each type of equipment but also allow the selection of equipment, i.e. the inclusion or exclusion of different types of equipment. This makes it necessary to have decision variables about the part routing/operation assignment as well as for the CNC-machines / load-unload stations / transportation system selection and possibly also for pallet decisions implemented.

Part and equipment optimization: Besides equipment optimization, the part types to produce on the system can also be selected in an optimal way. Here the same decision variables as for equipment optimization are required. However, part routing/operation assignments are now extended in such a way that part selection is possible.

System and equipment optimization: Here the type and number of production systems necessary to fulfil production requirements, and also their equipment are selected. The models of this class are therefore also applicable in the system planning phase. Here all three of the above-listed categories of decision variables are applicable.

Part, system and equipment optimization: These models comprise all of the above-mentioned decisions. For that reason it can be considered as being identical to the issue of the basic planning stage.

The classes of optimization models described above can be positioned as decision models in a hierarchical tree comprising an upper class and several lower classes according to the concept introduced by Schneeweiß.[143] The upper class is given by the issue of the basic planning stage, i.e. a part, system and equipment decision model (see fig.7.4).

---

143   Schneeweiß, C.: Construction and selection of quantitative planning models - a general procedure illustrated with models for stock control, in: EJOR, 6(1981), pp.372-379; Schneeweiß, C.: Elemente einer Theorie betriebswirtschaftlicher Modellbildung, in: ZfB, 54(1984)5, pp.480-504

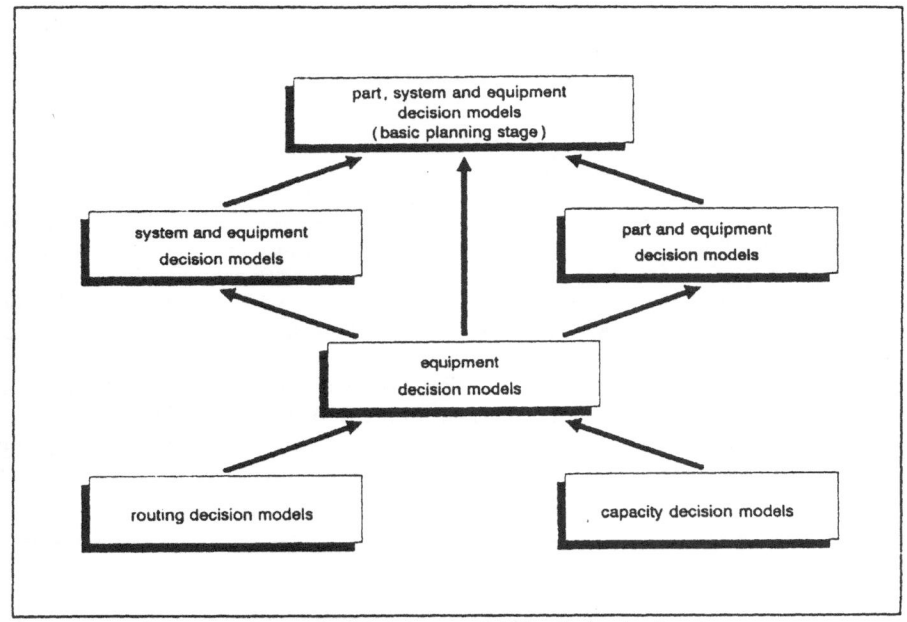

fig. 7.4: A hierarchy of decision models for the basic planning stage

7.5 A descriptive cost model for flexible manufacturing systems

In this chapter a cost model for flexible manufacturing systems is proposed. It is applicable for optimization models which, if productivity is optimized, minimizes costs (input) under the constraint of given production requirements (output).

When costs are considered, the flexible manufacturing system is configured according to a single, average period. Alternatively, as mentioned in chapter 7.3, changing levels in production requirements might require a multi-period model. Then, instead of cost considerations, a comparison of different possible configurations must be based on the present value of cash outflows over all periods. This is due to the fact that different alternative timings for the investment are then possible and must be compared, e.g. an investment in some overcapacity at the beginning versus a investment later in an expansion of the system.

Note that from the point of view of investment theory the latter approach with the consideration of present values within a multi-period model is more exact. However, if different timings need not to be compared, simplification, by only considering costs for an average period, is possible.

In chapter 8 a few models will be presented which relate to the following descriptive cost model. Furthermore one multi-period model is shown, which uses instead of costs the present value of cash outflows.

The cost model given below refers to the equipment selection problem. Here costs for the input of the flexible manufacturing system, which are the same for all possible configurations (e.g. raw material cost), can be neglected and thus only two major groups of costs are distinguished (fig.7.5):

- system costs
- operating costs.

This distinction can be based on the different cost behavior patterns as a function of the system's output. System costs can be considered as increasing in steps with increasing output of parts. This is because capacities of system resources are only available in discrete increments. However, operating costs continuously increase with output. Depending on how many parts are produced per period, operating costs increase accordingly. As plausible as this distinction is, it has one drawback. In the presented model of the causal relationship between parts and the production system of chapter 4.2.1, the tools were already shown as the link between an alternative process plan with its operations and the machine tools. As a result tool costs take a position in between system costs and operating costs. The question arises what kind of a cost function do tool costs have for different outputs.

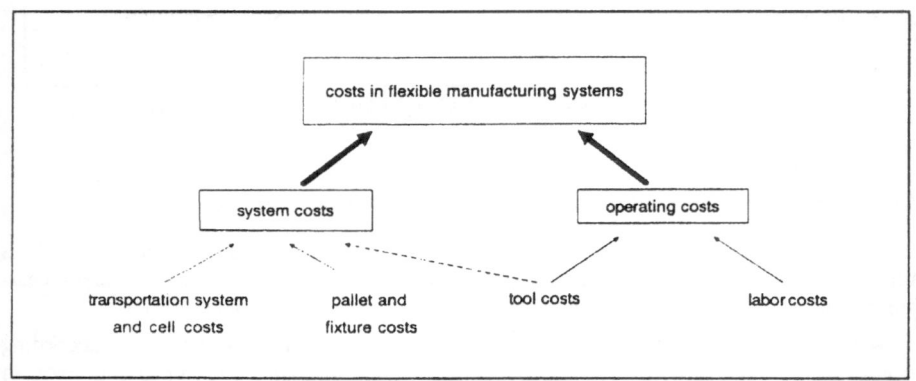

fig. 7.5: Cost model for the design process

If costs for wear during use are not negligible then tool costs must be regarded as increasing continuously in line with output. If however costs for wear during use are negligible, tool costs might be considered as independent of output. On the other hand, the necessary tools are determined by the process plans used for the production of the part types. Thus now tool costs depend on the output, but it is the qualitative structure and not the quantity of output on which they are dependent. To solve this issue it is suggested that tool costs should be treated as continuously increasing operating costs by evaluating them approximately through the relation of the operation time to the average expected lifetime of the tool.

System costs are costs concerning the transportation system, the cells, and the pallets. For each possible piece of equipment ($\forall$ m ϵ Conf., with m = 1 for the transportation system) linear increasing or just general increasing cost functions are assumed. The transportation system costs per period are a function of the number of vehicles y_1, the cell costs per period of the number of identical CNC-machines (load/unload stations) y_m, and the pallet costs per period of the number of pallets N in the system

Linear cost functions for cell costs and the transportation system are described by the following formulations:

$IF_1 + IV_1 \cdot y_1$

```
      ⌐
      |    ⌐ number of carriers
      |    
      |    ⌐ costs per period for one carrier
      ⌐
      ⌐ fixed costs per period for the transportation system
```

$IF_m + IV_m \cdot y_m$ $m = \{2,3,4...,M\}$

```
      ⌐
      |    ⌐ number of CNC-machines (load/unload station) in cell m
      |    
      |    ⌐ costs per period per CNC-machine (set-up table)
      ⌐
      ⌐ fixed costs per period for cell m
```

Fixed cell costs or transportation system costs, i.e. costs independent of the number of machines, carriers or load/unload stations, can, for example, comprise costs for the control system, a part handling system, if required, or a local tool storage and handling system.

Alternatively cell costs and transportation system costs can also be described more generally by increasing costs functions using binary variables:

$\sum_b I_{b1} \cdot z_{b1}$

```
      ⌐
      |    ⌐
      |    |   ⌐ binary variable, which is one if b carriers are selected for the
      |    |     transportation system and zero if not
      |    
      |    ⌐ costs per period for the transportation system with b carriers
```

$\sum_b I_{bm} \cdot z_{bm}$ $m = \{2,3,4...,M\}$

```
      ⌐
      |    ⌐
      |    |   ⌐ binary variable, which is one if b machines are selected for the
      |    |     cell m and zero if not
      |    
      |    ⌐ costs per period for a cell with b machines, load/unload stations etc.
```

To the costs per pallet IP the costs for fixtures IFI_p for part type p can be added. If universal pallets are used, the average number of pallets with part type p in the system and thus of fixtures for part type p is given by the product of the average number of pallets N in the system with the fraction $frac_p$, i.e. the fraction each part type p has of the whole production. Also in-process inventory costs, if relevant, can easily be incorporated. They consist of capital costs CC_p of part type p multiplied by the average number of pallets of part type p, i.e. with $N \cdot frac_p$.

$$[IP + \sum_{p=1}^{P} (IFI_p + CC_p)] \cdot N \cdot frac_p$$

- average number of pallets with part type p in the system
- capital costs for part type p
- fixture costs per period for part type p
- costs per period for one pallet

If special pallets for each part type p are in the system, the costs per pallet, fixture costs and in-process inventory costs have the following form:

$$(IP_p + IFI_p + CC_p) \cdot N_p$$

- average number of pallets for part type p in the system
- capital costs for part type p
- fixture costs per period for part type p
- costs per period for one pallet for part type p

Operating costs are costs which can be assigned to the operations performed on a part of type p at a machine of type m due to the given process plan used by route r through the system. They can be further subdivided into variable tool costs and variable labor costs.

$$CO_{mpr} = CT_{mpr} + CL_{mpr}$$

- labor costs
- tool costs
- operating costs of part type p at machine m if route r is used

Thus, if q_{pr} is the fraction of the system's part flow for part type p on route r with

$$\sum_p \sum_r q_{pr} = 1$$

the average operating costs for producing one part in the system are given by:

$$\sum_p \sum_r \sum_m CO_{mpr} \cdot q_{pr}$$

- fraction of part flow on route r for part p
- operating costs of part type p at machine m if route r is used

8. Models for the optimal design of flexible manufacturing systems

In this chapter models applicable during the basic planning stage for the optimal design of flexible manufacturing systems are studied. An overview of the models already existing in literature is presented together with new models developed by the author.

According to the two classification criteria in chapter 7.2 and 7.4, the optimization models of the basic planning stage can be positioned in a two dimensional matrix. This matrix consists of two columns, one for models with an unlimited number of pallets [UP], i.e. static system behavior modelling, the other one for those with a limited number of pallets [LP], i.e. dynamic system behavior modelling (see fig.8.1). Furthermore, the rows are assigned to different classes, derived from the decision variables used [RO, CA, EQ, PAEQ, SYEQ, PASYEQ]. Thus each position in the matrix represents a certain combination of the two classification criteria. According to its position in this matrix, each model has a name showing its position and the author's name. If there are several models from the same author at one position a number in brackets is added.

system behavior → ↓ decision problems	unlimited number of pallets (static system behavior modelling) [UP]	limited number of pallets (dynamic system behavior modelling) [LP]
routing optimization [RO]	ROUP-Secco-Suardo -Kimemia/Cershwin -Avonts et al.	ROLP-Kimemia/Cershwin -Yao/Shanthikumar
capacity optimization [CA]		CALP-Vinod/Solberg -Dallery/Frein -Yao/Shanthikumar-(1) -Yao/Shanthikumar-(2) -Yao/Shanthikumar-(3) -Solot
equipment optimization [EQ]	EQUP-Graves/Whitney -Graves/Lamar	
part and equipment optimization [PAEQ]	PAEQUP-Whitney/Suri	
system and equipment optimization [SYEQ]	SYEQUP-Sarin/Chen	
part, system and equipment optimization [PASYEQ]		

fig. 8.1: Model matrix

The model matrix illustrates the primary direction of research for this dissertation. If models with a limited number of pallets are examined more closely, it can be observed that, due to their complexity, so far only models for routing or capacity optimization have been created. Thus our main objective is to provide an extension by developing models for equipment optimization with a limited number of pallets. To date no models are known in this area. However, to be able to solve models for this class, some new models for routing optimization with a limited number of pallets and equipment optimization with an unlimited number of pallets are developed. Hence as a by-product some improved models are obtained for other model classes.

By analyzing the models given in literature additional scope for improvement was found. Thus a new model for capacity optimization with a limited number of pallets considering a budget constraint, and equipment optimization with an unlimited number of pallets allowing changing levels in production demand are presented.

In the following chapters all of the above-listed models will be described. Each description starts with a general characterization of the model. Then its mathematical formulation is outlined, followed by an explanation of the solution procedure(s) used. Furthermore, for newly developed models, numerical examples are given. Finally each description ends with a discussion on the model's capabilities and limitations.

8.1 Models for routing optimization

Models for routing optimization assume a predefined design of a flexible manufacturing system. Their aim is to optimize the assignment of production requirements to alternative routes or to operations at alternative machines in such a way that a given objective is optimized while capacity constraints for the cells and the transportation system are respected.

8.1.1 Models with an unlimited number of pallets

As already outlined in chapter 7.2, models with an unlimited number of pallets neglect the dynamic system behavior. However, through linearization they allow simpler models to be designed. They are quite accurate if a deterministic, steady part flow is given with only a few part groups and almost no routing interference, i.e. with flow line character, or if the number of pallets in the system will be large.

8.1.1.1 Throughput maximization by Secco-Suardo

The model by Secco-Suardo performs routing optimization with the objective to maximize the throughput of a given flexible manufacturing system.[144] This is done by the allocation of production requirements of each part type p to a given set of alternative routes R_p through the system. Each route consists of a set of workloads requiring certain machine types, load/unload stations etc. for production. For example, one part has one process plan by which it is first treated at machine A, then at machine B and afterwards at machine C. Additionally it has a second alternative process plan, by which it visits machine A, then machine C and finally machine D. Note however, that the order in which the machines are visited is here of no relevance.

Model ROUP-Secco-Suardo

Indices:

m	:	index for cells or transportation systems
p	:	part type index
r	:	route index

---

144   Secco-Suardo, G.: Optimization of closed queueing networks, in: Complex Materials Handling and Assembly Systems Final Report Vol. III No. ESL-FR-834-3, Electr. Syst. Lab. M.I.T., Cambridge MA, July 1978, pp.31-32; Secco-Suardo, G.: Workload Optimization in a FMS Modelled as a Closed Network of Queues, in: Annals of the CIRP, 28(1979)1, p.382

Decision variable:

x_{pr} : production rate of part p using route r

Parameters:

$frac_p$: product rate, i.e. the fraction part p has of total production flow
s_m : servers (machines, load/unload stations etc.) at cell (transportation system) m
W_{mpr} : workload of part type p on route r at cell m
$T(\underline{x})$: throughput of the system

Objective function:

$$\max \; T(\underline{x}) = \sum_{p=1}^{P} \sum_{r=1}^{R_p} x_{pr} \qquad\qquad (8.1.1.1.1)$$

Constraints:

$$\sum_{r=1}^{R_p} x_{pr} = frac_p \cdot \underbrace{\sum_{o=1}^{P} \sum_{r=1}^{R_p} x_{or}}_{\text{production rate of the whole system}} \qquad \forall \; p \qquad (8.1.1.1.2)$$

$$\sum_{p=1}^{P} \sum_{r=1}^{R_p} x_{pr} \cdot W_{mpr} \leq s_m \qquad \forall \; m \qquad\qquad (8.1.1.1.3)$$

$$x_{pr} \geq 0 \qquad\qquad \forall \; p,r \qquad\qquad (8.1.1.1.4)$$

In the objective function the production rate over all routes r of all part types p is maximized. Constraint set (8.1.1.1.2) ensures, that the production rate of one part (left-hand side of the equation) equals the production rate of the whole system multiplied by the factor $frac_p$. This factor, the product rate, describes the given fixed fraction that part type p has of total production. The second constraint set (8.1.1.1.3) ensures that the workload does not exceed the capacity of the cell. This is done by multiplying the production rates x_{pr} by the workloads W_{mpr} at cell m. The summation over all routes r and part types p must be less than or equal to the number of servers s_m at cell m.

Procedure for solution

The model presented has a linear programming formulation. Hence it can be solved with the simplex algorithm.

Discussion

Input:

- Different alternative routes must be provided.

Model:

- Static system behavior modelling restricts the application of this model to flexible transfer lines and multi-lines or to flexible manufacturing systems with a large number of pallets, where costs dependent on the number of pallets are negligible.

- Possible machine breakdowns are not explicitly included in the model. However, the right-hand side of the capacity constraint may be adjusted downward by an availability factor to reflect the expected annual downtime for the machine.

Procedure:

- The optimal solution is obtained.
- Standard software for linear programming can be applied.

8.1.1.2 Throughput maximization by Kimemia and Gershwin

The linear programming model by Kimemia and Gershwin is very similar to that of Secco-Suardo in the previous chapter. Again the goal is to maximize the throughput of a given flexible manufacturing system. The difference between the two can be seen in the decision variable. Compared to the model by Secco-Suardo there are no routes but operations given and the algorithm works on a more basic level by generating routes by selecting the optimal machines for an operation.[145]

Model ROUP-Kimemia/Gershwin

Indices:

k	:	operation index
m	:	machine index
p	:	part type index

Decision variables:

x_{mpk} : throughput at machine m of operation k for part type p

Parameters:

$frac_p$	:	product rate, i.e. the fraction part p has of total production flow
t_{mpk}	:	processing time for part p at machine m for operation k

Objective function:

$$\max \sum_{m=1}^{M} \sum_{p=1}^{P} x_{mp1} \tag{8.1.1.2.1}$$

Constraints:

$$\sum_{m=1}^{M} x_{mpk} = frac_p \cdot \sum_{m=1}^{M} \sum_{o=1}^{P} x_{mok} \qquad \forall\ p,k \tag{8.1.1.2.2}$$

---

145   Kimemia, J.G., Gershwin, S.B.: Multicommodity Network Flow Optimization in Flexible Manufacturing Systems, in: Complex Materials Handling and Assembly Systems Final Report Vol. II No. ESL-FR-834-2, Electr. Syst. Lab. M.I.T., Cambridge MA, July 1978, pp.35-37; Kimemia, J.G., Gershwin, S.B.: Network Flow Optimization in Flexible Manufacturing Systems, in: Proc. IEEE Conf. on Decision and Control 1979 pp.633-639; Kimemia, J.G., Gershwin, S.B.: Flow Optimization in Flexible Manufacturing Systems, in: IJPR 23(1985)1, pp.81-96

$$\sum_{m=1}^{M} x_{mpk+1} = \sum_{m=1}^{M} x_{mpk} \qquad \forall \, p, \; k=\{1,\ldots,K_p-1\} \qquad\qquad (8.1.1.2.3)$$

$$\sum_{p=1}^{P} \sum_{k=1}^{K} x_{mpk} \cdot t_{mpk} \leq 1 \qquad \forall \, m \qquad\qquad (8.1.1.2.4)$$

$$x_{mpk} \geq 0 \qquad \forall \, m,p,k \qquad\qquad (8.1.1.2.5)$$

In the objective function the throughput x_{mp1} for the first operation is maximized over all machines m and all parts p. Based on the constraint set (8.1.1.2.3) for continuous flow, it is then ensured that the throughput optimization for all the other remaining operations is the same as for the first operation. The two constraint sets (8.1.1.2.2) and (8.1.1.2.4) are very similar to those of the model by Secco-Suardo. Constraint set (8.1.1.2.2) fixes the fraction $frac_p$ of part type p of total production, whereas constraint set (8.1.1.2.4) takes care of capacity limitations. The production time for an operation k of a part type p at machine m is given by t_{mpk}. Note, that in this formulation of Kimemia and Gershwin the number of servers at each cell is limited to one. By replacing the 1 with the number of machines, the model can be easily extended to incorporate also cells with several machines.

Procedure for solution

The model presented has a linear programming formulation and can be solved with the simplex algorithm.

Discussion

Input:
- Compared with the model by Secco-Suardo only operations and their possible assignment to different machines must be provided.

Model:
- Static system behavior modelling restricts the application of this model to flexible transfer lines and multi-lines or to flexible manufacturing systems with a large number of pallets, where costs dependent on the number of pallets are negligible.
- Possible machine breakdowns are not explicitly included in the model. However, the right-hand side of the capacity constraint may be adjusted downward by an availability factor to reflect the expected annual downtime for the machine.
- Another limitation is caused by the consideration of single operations. If the transportation time is not very small compared to the processing times at the machines, they cannot be neglected and have to be incorporated. This can be easily done by considering transportation as a normal operation. However, in this case savings due to the assignment of successive operations to one cell cannot be identified.

Procedure:
- The optimal solution is obtained.

- Standard software for linear programming can be applied.

8.1.1.3 Model by Avonts et al.

In the model formulation by Avonts et al. the utilization of the flexible manufacturing system is maximized. This is achieved by assigning the production flow on different routes so that a given production demand for part type p is not exceeded.[146]

Model ROUP-Avonts et al.

Indices:
p	:	part type index
m	:	machine index
r	:	route index

Decision variable:
x_{pr} :   production flow for part p on route r

Parameters:
K_m	:	available capacity of machine type m
R_{pmax}	:	production demand of part type p
W_{mpr}	:	workload for part p at machine m on route r

Objective function

$$\max \sum_{m=1}^{M} \sum_{p=1}^{P} \sum_{r=1}^{R_p} W_{mpr} \cdot x_{pr} \qquad (8.1.1.3.1)$$

Constraints

$$\sum_{p=1}^{P} \sum_{r=1}^{R_p} W_{mpr} \cdot x_{pr} \leq K_m \quad \forall\ m \qquad (8.1.1.3.2)$$

$$\sum_{r=1}^{R_p} x_{pr} \leq R_{pmax} \quad \forall\ p \qquad (8.1.1.3.3)$$

$$x_{pr} \geq 0 \quad \forall\ p,r \qquad (8.1.1.3.4)$$

In the objective function (8.1.1.3.1) the summation over all production flows x_{pr} of part type p on route r multiplied by their workloads W_{mpr} on machine m yields the (unnormalized) utilization of the system. In the first constraint set (8.1.1.3.2) the utilization is limited to the available capacity K_m of each machine m. To ensure that the production demand R_{pmax} of part type p is not exceeded, in (8.1.1.3.3) the production flow x_{pr} of part p over all routes $r = \{1,..,R_p\}$ is less or equal to R_{pmax}.

---

146   Avonts, L.H., Gelders, L.F., Wassenhove Van, L.N.: Allocation work between an FMS and a conventional jobshop: A case study, in: EJOR, 33(1988), pp.245-256

Procedure for solution

The model presented has a linear programming formulation and can be solved with the simplex algorithm.

Discussion

Input:
- Different alternative routes must be provided.

Model:
- The model was originally designed for operational planning purposes and is only mentioned here for completeness. For the design of a flexible manufacturing system optimization of the utilization is not very helpful, because low cost and high throughput considerations are the major objectives during the design process.[147] For example, if machines are selected for a system so that a high utilization of the invested capital is achieved, it is better to have a particulary high utilization of expensive machines. Therefore cost-weighted utilization might be more appropriate. Only if the system already exists, an efficient use of the investment and therefore high utilization of the existing resources is desired.
- Static system behavior modelling restricts the application of this model to flexible manufacturing systems with flow-line character, i.e. to flexible transfer lines and -multi-lines or to flexible manufacturing systems with a large number of pallets, where costs dependent on the number of pallets are negligible.
- Possible machine breakdowns are not explicitly included in the model. However, the right-hand side of the capacity constraint may be adjusted downward by an availability factor to reflect the expected annual downtime for the machine.

Procedure:
- The optimal solution is obtained.
- Standard software for linear programming can be applied.

8.1.2 Models with a limited number of pallets

Models with a limited number of pallets incorporate queueing theory to allow the consideration of dynamic system behavior. The more variable and flexible the part flows in systems are, the better they can be analyzed by stochastic queueing models. However, this advantage is accompanied by increasing difficulties with the models' solution procedures.

8.1.2.1 Throughput maximization by Kimemia and Gershwin

Kimemia and Gershwin developed several models for flow optimization, of which a

---

147   see chapter 4.2.2

linear formulation was already presented in chapter 8.1.1.2. Here an extended version of the former model is shown, considering dynamic system behavior by referring to queueing theory.[148] In the model the flow rate x_{mpk} of a type p piece at workstation m for operation k is optimized, so that total throughput is maximum. This is done subject to a constraint imposed on the average level of in-process inventory, which is required to be less than or equal to a given value N.

Model ROLP-Kimemia/Gershwin

Indices:

k	:	operation index
m	:	index for machines or arcs
p	:	part type index

Decision variables:

x_{mpk}	:	flow rate of part type p to machine m for operation k

Parameters:

K_m	:	capacity of arc m
$frac_p$	:	product rate, i.e. the fraction part p has of total production flow
N	:	maximal number of parts in system (inprocess inventory)
$\bar{n}_m(\underline{x})$	:	average number of parts in queue and under process at machine m
r_{mp}	:	flow rate of type p pieces on arc m of the network
t_m	:	average travel time on arc m
t_{mpk}	:	processing time for part p at machine m for operation k
U_m	:	utilization of machine m

Objective function:

$$\max \quad \sum_{p=1}^{P} x_{1p1} \tag{8.1.2.1.1}$$

Constraints:

$$\sum_{m=1}^{M} x_{mpk+1} - \sum_{m=1}^{M} x_{mpk} = 0 \qquad \forall\ p,\ k=\{1,\ldots,K_p-1\} \tag{8.1.2.1.2}$$

$$\sum_{m=1}^{M} x_{mpk} - \frac{frac_p}{frac_1} \cdot \sum_{m=1}^{M} x_{m1k} = 0 \quad \forall\ p \tag{8.1.2.1.3}$$

---

148   The model presented is an essence out of two publications on the model: Kimemia, J.G., Gershwin, S.B.: Multicommodity Network Flow Optimization in Flexible Manufacturing Systems, in: Complex Materials Handling and Assembly Systems Final Report Vol. II No. ESL-FR-834-2, Electr. Syst. Lab. M.I.T., Cambridge MA, July 1978, pp.31-32; Kimemia, J.G., Gershwin, S.B.: Network Flow Optimization in Flexible Manufacturing Systems, in: Proc. IEEE Conf. on Decision and Control 1979 pp.634-635; Note that in the second paper the fixation of the production rate by constraint sets (2.2) and (2.4) has to be eliminated if our aim is to maximize total production in the objective function (2.10). It is only necessary to fix the relative amounts of production between different parts (see constraint (2.31) or (2.28) in the first paper). Further the continuity of operational flow must be ensured and is missing in the second paper (see constraint set (2.27) in the first paper).

$$U_m = \sum_{p=1}^{P} \sum_{k=1}^{K_p} x_{mpk} \cdot t_{mpk} \leq 1 \qquad m=\{1,\ldots,M\} \tag{8.1.2.1.4}$$

$$\sum_{n \in A(m)} r_{np} = \sum_{k=1}^{K} x_{mpk} \qquad m=\{1,\ldots,M\} \tag{8.1.2.1.5}$$

$$\sum_{n \in A(m)} r_{np} - \sum_{o \in D(m)} r_{op} = 0 \qquad \forall\ m{>}M \tag{8.1.2.1.6}$$

$$\sum_{p=1}^{P} r_{mp} \leq K_m \qquad \forall\ m{>}M \tag{8.1.2.1.7}$$

$$\sum_{m=1}^{M} \bar{n}_m(\underline{x}) + \sum_{p} \sum_{m{>}M} t_m \cdot r_{mp} \leq N \tag{8.1.2.1.8}$$

$$x_{mpk} \geq 0 \qquad \forall\ m,p,k \qquad r_{mp} \geq 0 \qquad \forall\ m,p \tag{8.1.2.1.9}$$

In the objective function total production of the system is optimized by maximizing the total flow rate x_{1p1} of all parts p at the loading station 1 performing their first operation. This is due to the first constraint set (8.1.2.1.2), which ensures continuity of operation flow, i.e. the production rate of part type p remains constant over all operations k.

The constraint set (8.1.2.1.3) fixes the relation between the production of different part types p. Furthermore the fraction that one part type p has of total production is fixed by $frac_p$. Hence we have

$$\sum_{p=1}^{P} frac_p = 1.$$

The limitation in capacity of a workstation is given by constraint set (8.1.2.1.4). The utilization U_m given by the summation of all multiplicative terms of the flow rate x_{mpk} with the operation time t_{mpk} at that station m can not exceed 1.

The transportation system in this model is designed as a network of nodes and arcs. Thereby the nodes are either merges or diverges of arcs, or the actual workstations themselves. For convenience, it is assumed that the nodes are numbered so that the first M are workstations and the remainder merges or diverges. The loading and unloading stations are labelled 1 and M so that operation 1 becomes the loading operation and K_p the unloading operation for a type p piece. The flow rate on one arc m of the transportation system for a type p piece is given by r_{mp}. To ensure the conservation of flow, in constraint set (8.1.2.1.6) the flow of the set A(m) of arcs leading to node m must be equal to the flow of set D(m) of arcs carrying pieces away from node m. Furthermore we can state in (8.1.2.1.5) that the flow of type p pieces to (from) station m of the set A(m) (D(m)) is equal to the flow rate x_{mpk} of all operations for part type p done at station m. To reflect the fact that each arc has only a limited capacity K_m constraint set (8.1.2.1.7) is given. Here the sum over the flow rates r_{mp} of all type p pieces is not allowed to exceed K_m.

Finally the number of parts in the system must be less than or equal to N. This number is obtained by the summation of the average queue lengths $\bar{n}_m(\underline{x})$ at all the stations and of all parts in the transportation system. The latter is calculated by the multiplication of the flow rates r_{mp} with the average travel time t_m on each arc with $m > M$. The queue length $\bar{n}_m(\underline{x})$ at each workstation can be evaluated by applying queueing theory. If service times can be assumed to be exponentially distributed and the arrival of pieces follow a Poisson process, Kimemia and Gershwin propose studying the workstations in isolation as M/M/1 queues.[149] Secco-Suardo has shown that this assumption is consistent with the network of queues approach with N being large.[150] In this case the average queue length is:

$$\bar{n}_m(\underline{x}) \;=\; \frac{U_m(\underline{x})}{1 - U_m(\underline{x})}$$

Procedure for solution

The method proposed by Kimemia and Gershwin involves breaking the problem into flow-generating linear programs and non-linear optimization problems with a reduced number of variables and simpler constraints.

The model is transferred from a nonlinear constrained problem into a linearly constrained problem by the formulation of an augmented Lagrangian objective function. By this means the nonlinear constraint set (8.1.2.1.8) is attached to the objective function. Then Gerla's EF-Method[151] (Extremal Flow Method) is applied, to iteratively solve a sequence of linearly constraint problems. The decomposition method of Dantzig and Wolfe is used to solve the flow generative subproblems.[152]

Discussion

Input:
- Only operations and their possible assignment to different machines must be provided.

Model:
- Dynamic system behavior modelling allows the application of this model to flexible machining systems.

- Studying the workstation in isolation as M/M/1 queues assumes a large number of pallets in the system.

- Only single-server cells and no multiple-server cells are considered. However, it can be

149   Kimemia, J.G., Gershwin, S.B.: Multicommodity Network Flow Optimization in Flexible Manufacturing Systems, in: Complex Materials Handling and Assembly Systems Final Report Vol. II No. ESL-FR-834-2, Electr. Syst. Lab. M.I.T., Cambridge MA, July 1978, p.25

150   Secco-Suardo, G.: Optimization of closed queueing networks, in: Complex Materials Handling and Assembly Systems Final Report Vol. III No. ESL-FR-834-3, Electr. Syst. Lab. M.I.T., Cambridge MA, July 1978, pp.21-22;

151   Gerla, M.: The design of store-and-forward (S/F) networks for computer communications, Ph.D. thesis Dep. Comp. Science, Univ. Calif. Los Angeles, 1973; Cantor, D.G., Gerla, M.: Optimal Routing in a Packet Switched Computer Network, in: IEEE Trans. on Computers, C-23(1974)10, pp.1062-1069;

152   Kimemia, J.G., Gershwin, S.B.: Multicommodity Network Flow Optimization in Flexible Manufacturing Systems, in: Complex Materials Handling and Assembly Systems Final Report Vol. II No. ESL-FR-834-2, Electr. Syst. Lab. M.I.T., Cambridge MA, July 1978, pp.47,51-63; Kimemia, J.G., Gershwin, S.B.: Network Flow Optimization in Flexible Manufacturing Systems, in: Proc. IEEE Conf. on Decision and Control 1979 pp.635-636;

extended by including also M/M/s queues in equilibrium.[153]

Procedure:
- The optimal solution is obtained.
- The procedure for the solution is quite complicated.

8.1.2.2 Throughput maximization by Yao and Shanthikumar

The model by Yao and Shanthikumar optimizes input rates to m manufacturing cells, so that the total throughput of the system is maximized. Each cell consists of s_m parallel servers without extra waiting room. An arriving part receives service if there are any servers available, and leaves the system when service is completed. Hence there is only one operation performed on an arriving part. If all servers are occupied, the arriving part is blocked from entry and transferred to an overflow system.[154]

Model ROLP-Yao/Shanthikumar

Index:
m : cell index

Decision variable:
x_m : input rate at cell m

Parameters:
$B(s_m, \rho_m)$: blocking probability
bl_m : limit for blocking probability at cell m
μ_m : service rate at cell m
n_m : equilibrium number of parts in cell m
R : production rate of the upstream production stage
s_m : number of servers at cell m
T_m : throughput of cell m

Objective function:

$$\max \sum_{m=1}^{M} T_m(x_m) \qquad (8.1.2.2.1)$$

Constraints:

$$B(s_m, \rho_m) \le bl_m \qquad \forall\, m \ , \ \text{and with } \rho_m = x_m\,/\,\mu_m \qquad (8.1.2.2.2)$$

153 Kimemia, J.G., Gershwin, S.B.: Multicommodity Network Flow Optimization in Flexible Manufacturing Systems, in: Complex Materials Handling and Assembly Systems Final Report Vol. II No. ESL-FR-834-2, Electr. Syst. Lab. M.I.T., Cambridge MA, July 1978, p.25

154 Yao, D.D., Shanthikumar, J.G.: The optimal input rates to a system of manufacturing cells, in: INFOR, 25(1987)1, pp.57-65; Yao, D.D., Shanthikumar, J.G.: Some resource allocation problems in multi-cell systems, in: Proc. 2nd ORSA/TIMS Conf. on Flexible Manufacturing Systems: Operations Research Models and Applications, edited by K.E. Stecke and R. Suri, Amsterdam 1986, pp.250-253

$$x_m \geq 0 \quad \forall \ m, \qquad |\underline{x}| = \sum_{m=1}^{M} x_m = R \qquad\qquad (8.1.2.2.3)$$

In the objective function the sum of cell throughputs T_m is maximized. The first constraint set limits the blocking probability $B(s_m, \rho_m)$ to a given value bl_m at each cell m. The last two constraint sets describe the input rate x_m. Its value has to be greater or equal to zero and the sum over all input rates yields the given constant R, which is the production rate of the upstream production stage generating the input to the system under consideration.[155]

Procedure for solution

First the above maximization model will be transformed to a minimization model applying queueing theory. Yao and Shanthikumar model each cell m having s_m servers as an Erlang loss queue of the type $M/G/s_m/s_m$. The throughput of a cell is therefore derived as follows:

$$T_m(x_m) = E[\mu_m n_m] = \mu_m \cdot E[n_m] = x_m \cdot P(s_m - 1, \rho_m)/P(s_m, \rho_m) \qquad (8.1.2.2.4)$$
$$= x_m \cdot [1 - B(s_m, \rho_m)]$$

$$\text{with } P(s_m, \rho_m) = \sum_{k=0}^{s_m} \rho_m^{\ k}/k! \qquad \text{and} \qquad \rho_m = x_m / \mu_m .$$

The mean service time of cell m is given by $1/\mu_m$. For the blocking probability we obtain:

$$B(s_m, \rho_m) = (\rho_m^{\ s_m}/s_m!) \ / \ (\sum_{k=0}^{s_m} \rho_m^{\ k}/k!)$$

With the help of (8.1.2.2.4) the objective function can be transformed to:

$$\sum_{m=1}^{M} T_m(x_m) = R - \sum_{m=1}^{M} x_m \cdot B(s_m, x_m/\mu_m)$$

The blocking constraints (8.1.2.2.2) are changed to $x_m \leq u_m \ \forall \ m$, where $\underline{u}$ is the solution of the following set of equations:

$$B(s_m, u_m/\mu_m) = bl_m \qquad \forall \ m$$

The maximization problem is then equivalent to the following minimization problem:

Objective function:

$$\min \ f(x_m) = \sum_{m=1}^{M} x_m \cdot B(s_m, x_m/\mu_m)$$

---

155   Yao, D.D., Shanthikumar, J.G.: Some resource allocation problems in multi-cell systems, in: Proc. 2nd ORSA/TIMS Conf. on Flexible Manufacturing Systems: Operations Research Models and Applications, edited by K.E. Stecke and R. Suri, Amsterdam 1986, p.250

Constraints:

$$u_m \geq x_m \geq 0 \quad \forall \; m, \qquad |\underline{x}| = \sum_{m=1}^{M} x_m = R$$

The model now consists of a convex function, which has to be minimized over a convex set. To solve the problem Yao and Shanthikumar propose a modified version of the Frank-Wolfe algorithm.[156]

Discussion

Input:
- The limit for the blocking probability bl_m at cell m must be provided.

Model:
- Yao and Shanthikumar model each cell as a multiple parallel server. Because the input rates are the decision variables, arriving parts to the system can be directed to any one of the given m cells and then leave the system. Therefore it can be concluded that for each part there is only one kind of operation performed in the whole system. Because parts are directed to different cells, only changes in service times for a part can occur. These facts strongly limit the application of the model to a special kind of flexible manufacturing system.

Procedure:
- The optimal solution is obtained.

8.1.2.3 A new model for delay minimization

The FDC algorithm (Flow Deviation algorithm for Closed queueing networks) was originally developed by Kobayashi and Gerla to calculate the optimal routing of jobs in computer systems.[157] It incorporates the closed queueing network model into the FD algorithm to minimize the overall average delay, i.e. the time an average part spends in the system, its throughput time. For closed queueing networks with a given number of parts in the system, this is, according to Little's law, equivalent to maximizing total average throughput.[158] Unfortunately this algorithm by Kobayashi and Gerla cannot directly be applied to flexible manufacturing systems, because the service rate at a station is considered to have a constant for its station characteristic value. Whereas for computer systems this assumption might be feasible, for flexible manufacturing systems this is not the case. In the latter a change in the part flow on alternative routes or a change of the part mix causes a change in the average service rate. Thus another decision variable is introduced by considering the fraction of part flows q_{pr} on each route r instead of the

156   for the proof of convexity of the model and a proof for convergence of the algorithm to the optimal solution see: Yao, D.D., Shanthikumar, J.G.: The optimal input rates to a system of manufacturing cells, in: INFOR, 25(1987)1, pp.57-65;
157   Kobayashi, H., Gerla, M.: Optimal Routing in Closed Queueing Networks, in: ACM Trans. Comp. Syst., 1(1983)4, pp.294-310; For a description of the FD-Algorithms, see chapter 6.1.1.2.
158   Little, J.D.C.: A Proof of the Queueing Formula L = λ · W, in: OR 9(1961), pp.383-387

relative flow e_m of parts, and the model is changed accordingly.[159]

Model ROLP-NEW-(1)

Indices:

m	:	index for cells or transportation system
p	:	index for part types
r	:	index for routes

Decision variables:

q_{pr}	:	fraction of part flow on route r for part p

Parameters:

a	:	Lagrange multiplier
D_N	:	overall average delay of a pallet in a system with N pallets, i.e. its throughput time
$frac_p$	:	product rate, i.e. the fraction part p has of total production flow
$\bar{n}_m$	:	average queue length at station m
l_{pr}	:	route length of part type p on route r
N	:	number of pallets in the system
t_{mpr}	:	processing time for part type p on route r at machine m
T_p	:	throughput of part p
v_{mpr}	:	number of visits for part type p on route r at machine m

Objective function:

$$\min_{\underline{q}} D_N(\underline{q}) \tag{8.1.2.3.1}$$

Constraints:

$$\sum_r q_{pr} = frac_p \quad \forall\ p \quad \text{and with} \quad \sum_p frac_p = 1 \tag{8.1.2.3.2}$$

$$q_{pr} \geq 0 \quad \forall\ p, r \tag{8.1.2.3.3}$$

For a given flexible manufacturing system with N pallets in the system the overall average delay $D_N(\underline{q})$, depending on the flow fractions $\underline{q}$ chosen for each route r of part type p is minimized. Alternatively the throughput function could be maximized, because by Little's law those two functions are reciprocally interrelated. However, only for the delay function convexity can be proved (see proof in Appendix A). Since the reciprocal function of a convex function is not necessarily concave, Stecke has in fact shown nonconcavity for certain cases of the throughput function.[160] Thus to obtain a convex model formulation, the delay function is minimized in the objective function.

In the first constraint set the sum over all flow fractions q_{pr} of all alternative routes r of each part type p is fixed to the product rate $frac_p$. Note that the sum of all product rates $frac_p$ is equal to one, because they describe the fraction each part type p has of total production. In the last constraint set the non-negativity of flows is stated.

---

159   See definitions of workloads in chapter 6.1.2.2.2

160   Stecke, K.E.: On the Nonconcavity of Throughput in Certain Closed Queueing Networks, in: Performance Evaluation, 6(1986), pp.293-305

Procedure for solution

Before the algorithm itself is described, a few introductory remarks will be given.

First the sensitivity of the average delay function $D_N(\underline{q})$ with regard to the flow fraction q_{pr} will be derived. By applying the chain rule, it can be stated that:

$$\frac{\partial D_N(\underline{q})}{\partial q_{pr}} = \sum_m \frac{\partial D_N(\underline{q})}{\partial W_m} \cdot \frac{\partial W_m}{\partial q_{pr}}$$

The sensitivity of the average delay function $D_N(\underline{r})$ with regard to the relative workload W_m at the (load dependent) station m can be derived form standard queueing network theory and can be written in accordance to Gordon and Dowdy as:[161]

$$\frac{\partial D_N(\underline{q})}{\partial W_m} = \frac{D_N(\underline{q})}{W_m} \cdot (\bar{n}_m(N) - \bar{n}_m(N-1))$$

The variable $\bar{n}_m$ is the average queue length at station m, if N or respectively N-1 pallets are in the system. If the relative workload W_m are defined as:[162]

$$W_m = \sum_p \sum_r t_{mpr} \cdot v_{mpr} \cdot q_{pr}$$

its derivation with regard to q_{pr} can be stated as:

$$\frac{\partial W_m}{\partial q_{pr}} = t_{mpr} \cdot v_{mpr}$$

As a result the sensitivity of the average delay function $D_N(\underline{q})$ with regard to the flow fraction q_{pr} is given by:

$$\frac{\partial D_N(\underline{q})}{\partial q_{pr}} = \sum_m \frac{D_N(\underline{q})}{\sum_k \sum_l t_{mkl} \cdot v_{mkl} \cdot q_{kl}} \cdot (\bar{n}_m(N) - \bar{n}_m(N-1)) \cdot t_{mpr} \cdot v_{mpr}$$

Now to each route r of part type p a route length l_{pr} is assigned as follows:

$$l_{pr} = \frac{\dfrac{\partial D_N(\underline{q})}{\partial q_{pr}}}{D_N(\underline{q})} = \sum_m \frac{(\bar{n}_m(N) - \bar{n}_m(N-1)) \cdot t_{mpr} \cdot v_{mpr}}{\sum_k \sum_l t_{mkl} \cdot v_{mkl} \cdot q_{kl}}$$

The route length l_{pr} expresses the relative change of the delay function $D_N(\underline{q})$ with respect to q_{pr}. Thus with the help of the route length l_{pr} the marginal delay for route q_{pr} is given by $l_{pr} \cdot D_N(\underline{q})$; i.e. a reassignment of part flows from flow vector $\underline{q}$ to $\underline{q}'$ yields a change in average delay $\Delta D_N(\underline{q})$ of:

$$\Delta D_N(\underline{q}) \approx \frac{\partial D_N(\underline{q})}{\partial \underline{q}} \cdot \Delta \underline{q} = D_N(\underline{q}) \cdot \sum_p \sum_r l_{pr} \cdot (q'_{pr} - q_{pr})$$

---

161   Gordon, K.D., Dowdy, L.W.: The Impact of Certain Parameter Estimation Errors in Queueing Network Models, in: ACM Perf. Eval. Rev., 2(May 1980) pp.3-9

162   see chapter 6.1.2.2.2 for the different definitions of relative workloads

To ensure that the global maximum can be found, convexity of the delay function must be given.

Theorem 2: The delay function $D_N(\underline{q})$ is convex in $\underline{q}$.

Proof: see Appendix A

With a convex delay function $D_N(\underline{q})$ the conditions for the minimum of the delay function can be stated.[163] They can be derived with the help of the Kuhn-Tucker conditions. First the Lagrange function $LF(\underline{q},\underline{\alpha})$ is formed:

$$LF(\underline{q},\underline{\alpha}) = D_N(\underline{q}) - \sum_p \alpha_p \cdot (\sum_r q_{pr} - frac_p)$$

From the Kuhn-Tucker conditions[164]:

$$\frac{\partial LF(\underline{q},\underline{\alpha})}{\partial q_{pr}} \cdot q_{mj} = \frac{\partial D_N(\underline{q})}{\partial q_{pr}} \cdot q_{pr} - \alpha_p \cdot q_{pr} = 0$$

$$\frac{\partial LF(\underline{q},\underline{\alpha})}{\partial q_{pr}} = \frac{\partial D_N(\underline{q})}{\partial q_{pr}} - \alpha_p \geq 0$$

the following conditions are obtained:

$$\forall r \ \text{with} \ q_{pr}>0 \qquad \frac{\partial D_N(\underline{q})}{\partial q_{pr}} = l_{pr} \cdot D_N(\underline{q}) = \alpha_p$$

$$\forall r \ \text{with} \ q_{pr}=0 \qquad \frac{\partial D_N(\underline{q})}{\partial q_{pr}} = l_{pr} \cdot D_N(\underline{q}) \geq \alpha_p$$

An intuitive justification of the above result is provided by the fact that at equilibrium all routes in use, i.e. $q_{pr} > 0$, must have the same marginal delay. A nonuniform distribution of route lengths can always be equalized with consequent reduction in delay because of convexity. If for example, two nonnegative flows q_{p1} and q_{p2} for part type p have the route length $l_{p1} > l_{p2}$, then a shift ΔS of flow from q_{p1} to q_{p2} causes a change in delay $\Delta D_N(\underline{q})$ given by:

$$\Delta D_N(\underline{q}) = \Delta S \cdot D_N(\underline{q}) \cdot (l_{p2} - l_{p1}) < 0$$

Consequently an improving step consists in shifting traffic to the shortest path as long as a nonuniform distribution of route length for nonnegative flows exists, i.e. a certain tolerance level hereof is exceeded.

Incorporating these observations into the FD algorithm, the following procedure is obtained:

---

163   Further results of concavity and convexity in Closed Queueing Networks theory are shown in chapter 6.1.2.2.3

164   Fiacco, A.V., McCormick, G.B.: Nonlinear Programming: Sequential Unconstrained Minimization Techniques, New York 1968, pp.19-20

Adapted FDC algorithm
Step 0:   find a feasible starting vector $\underline{q}^0$ set $n=0$
Step 1:   compute the route length l_{pr} ∀ p,r with the help of the CQN algorithm
Step 2:   solve the shortest route problem for each part p   define p x r flow matrix $\underline{y}^*$ having the values $frac_p$ at the shortest routes   and zeroes elsewhere.
Step 3:   compute the incremental delay: $$\sum_p \sum_r l_{pr} \cdot (q_{pr}{}^n - y_{pr}{}^*)$$
Step 4:   if the incremental delay is smaller than a properly chosen tolerance β with β>0: $$\left| \sum_p \sum_r l_{mj} \cdot (q_{pr}{}^n - y_{pr}{}^*) \right| < \beta \quad \text{stop;}$$ else go to step 5
Step 5:   solve $$\min_{0 \le \alpha \le 1} P[(1-\alpha) \cdot \underline{q}^n + \alpha^* \cdot \underline{y}^*]$$ with line search algorithm (golden section method or Fibonacci search)   to obtain α^*.   calculate $\underline{q}^{n+1} = (1 - \alpha^*) \cdot \underline{q}^n + \alpha^* \cdot \underline{y}^*$
Step 6:   set $n=n+1$ go to step 1

alg. 8.1: Adapted FDC algorithm

Example

Use of the above model is demonstrated by an example given below. It consists of a part family of 3 different part types {part1, part2, part3}. For each part type different, alternative routes exist, i.e. part1 and part2 have two, and part3 has three alternative routes. In tab.8.1 part data concerning the workloads on each route of the part types are given.

part	route	L/U St.	mill-1	mill-2	mill-3	vtl-1	vtl-2	transp.
part1	route1	10.00	15.00	30.00	-	-	-	9.00
	route2	15.00	20.00	-	-	20.00	-	9.00
part2	route1	5.00	-	-	10.00	5.00	16.00	12.00
	route2	5.00	20.00	-	-	20.00	-	9.00
part3	route1	15.00	10.00	5.00	5.00	10.00	-	15.00
	route2	15.00	-	10.00	30.00	-	5.00	12.00
	route3	10.00	-	30.00	30.00	-	-	9.00

tab. 8.1: Processing times at each machine for each route

In tab.8.2 the given configuration of the flexible manufacturing system is shown. It consists of a transportation system (transp.) with four vehicles, a cell of four load/unload stations (L/U St.), several cells with milling machines (mill-1, mill-2, mill-3), and with vertical turret lathes (vtl-1, vtl-2). Further the product rate, i.e. the fraction $frac_p$ each part type p has of total production, is presented in tab.8.3.

	L/U St.	mill-1	mill-2	mill-3	vtl-1	vtl-2	transp.
s_m	3	2	2	1	2	2	3

tab. 8.2: Configuration of the given flexible manufacturing system

	part1	part2	part3
$frac_p$	0.289	0.395	0.316

tab. 8.3: Product rates

As the result of the algorithm described above, the flow fractions with the resulting throughput T_{pr} of each route is obtained (see tab.8.4) for 20 pallets in the system.

	part1		part2		part3		
	route1	route2	route1	route2	route1	route2	route3
q_{pr}	0.212	0.078	0.319	0.075	0.316	–	–
T_{pr}	6839	2501	10305	2429	10188	–	–

tab. 8.4: Flow fraction on each route

Extension for specialized types of pallets:

The algorithm described above is valid for flexible manufacturing systems with one pallet type. Therefore it can only be applied to systems with universal pallets. If, however, specialized pallets are introduced into the system, a good evaluation of average delay and throughput can only be made through networks allowing several pallet types. This is due to the fact, that if a part has finished process and has to be replaced, only certain part types can be mounted on specialized pallets, whereas any part type can be mounted next on universal pallets. For that reason the algorithm is extended. A major drawback of this extension however is to be seen in the fact that, because of the possibility of multiple local minima, finding global optimality is not guaranteed. Heuristically, a search for the global optimum can be made by starting the algorithm with several different initial flows.[165]

Except for some small changes, the model formulation given in the previous chapter can be maintained. The index for part type p now is changed to c to account for a pallet of type c. Another difference can be seen in the delay function, which now depends on a pallet vector $\underline{N}$ instead of the number of universal pallets N. Otherwise the model formulation is analogous.

Procedure for solution

The route length l_{cr} can be derived in the same way as in the case of universal pallets. With the help of the sensitivity of the average delay function $D(\underline{N},\underline{q})$ with regard to the relative workload W_{mc} at station m for pallet type c [166]

$$\frac{\partial D(\underline{N},\underline{q})}{\partial W_{mc}} = D(\underline{N},\underline{q}) \cdot [T_c(\underline{N},\underline{q}) \cdot ts_{mc}(\underline{N},\underline{q}) - \sum_{c'} \frac{T_c \cdot (\underline{N},\underline{q})}{T(\underline{N},\underline{q})} \cdot T_c(\underline{N}-\underline{1}_{c'},\underline{q}) \cdot ts_{mc}(\underline{N}-\underline{1}_{c'},\underline{q})]$$

the route length l_{cr} is given by:

$$l_{cr} = \frac{\dfrac{\partial D(\underline{N},\underline{q})}{\partial q_{cr}}}{D(\underline{N},\underline{q})}$$

165 Kobayashi, H., Gerla, M.: Optimal Routing in Closed Queueing Networks, in: ACM Trans. Comp. Syst., 1(1983)4, p.303

166 Kobayashi, H., Gerla, M.: Optimal Routing in Closed Queueing Networks, in: ACM Trans. Comp. Syst., 1(1983)4, p.303

$$= \sum_m [T_c(\underline{N},g) \cdot ts_{mc}(\underline{N},g) - \sum_{c'} \frac{T_c \cdot (\underline{N},g)}{T(\underline{N},g)} \cdot T_c(\underline{N}-1_{c'},g) \cdot ts_{mc}(\underline{N}-1_{c'},g)] \cdot t_{mcr} \cdot v_{mcr}$$

The throughput of pallets of type c having $\underline{N}$ or respectively $\underline{N}-1_c$, pallets in the system is given by the variable T_c, whereas ts_{mc} accounts for the sojourn time for pallet type c at station m having $\underline{N}$ or respectively $\underline{N}-1_c$, pallets in the system. The service time t_{mcr} and number of visits v_{mcr} of pallet type c at station m on route r are also necessary to obtain the route length l_{cr}.

The solution procedure then performed with the modified route lengths is formally identical to the one for universal pallets, except that evaluations must now be done with a multi-chain queueing network algorithm for example PMVA.[167]

Discussion

Input:
- Different alternative routes must be provided.

Model:
- Dynamic system behavior modelling allows the application of this model to flexible machining systems.
- During the design phase of a flexible manufacturing system besides throughput maximization cost minimization is also a major objective.[168] The latter must be considered for route selection, when operating costs are quite different on alternative routes and when their amount has a considerable influence on whole costs for the system.
- Further limitations are given by the stochastic assumptions of closed queueing network theory.[169]

Procedure:
- Optimality can only be proven for the multi-server case with universal pallets.[170] When specialized pallets are considered, local optima are obtained.
- During each optimization step with a line search algorithm improvements are only made in the direction of one alternative route, i.e. the shortest route. Thus with an increasing number of alternative routes, computational requirements increase. Furthermore during line search, when a single class closed queueing network based on Buzen's algorithm is taken to evaluate the delay, the number of calculations at each step is given by $O(2 \cdot M \cdot N \cdot (N+1))$ to obtain $G(M,N)$.[171]

167　Shalev-Oren, S., Seidmann, A., Schweitzer, P.J.: Analysis of Flexible Manufacturing Systems with Priority Scheduling: PMVA, in: Annals of Operation Research, 3(1985), pp.115-139

168　see chapter 4.2.2

169　see chapter 6.1.2.2

170　see Appendix A

171　Buzen, J.P.: Computational Algorithms for Closed Queueing Networks with Exponential Servers, in: Comm. ACM, 16(1973)9, p.530

8.1.2.4 A new model for the minimization of routing costs

Whereas in the previous chapter a new model for throughput and delay optimization was proposed, this chapter presents a model for routing-cost optimization. The mathematical program minimizes operating costs for the production of a part family over alternative routes, while maintaining a minimum throughput R_{min}. For the design of flexible manufacturing systems these operating costs are relevant when they have besides system costs a considerable influence.[172]

Model ROLP-NEW-(2)

Indices:

m	:	index for cells, transportation system or assembly station
p	:	part type index
r	:	index for routes

Decision variables:

q_{pr}	:	fraction of part flow on route r for part p

Parameters:

α	:	Lagrange multiplier
CO_{mpr}	:	operating costs of part type p on route r at machine m
$frac_p$	:	product rate, i.e. the fraction part p has of total production flow
N	:	number of pallets in the system
R_{min}	:	required production rate for the system
T_N	:	system throughput with N pallets
w_1, w_2	:	penalty weights

Objective function:

$$\min_{\underline{q}} \quad \sum_m \sum_p \sum_r CO_{mpr} \cdot q_{pr} \qquad (8.1.2.4.1)$$

Constraints:

$$T_N(\underline{q}) \geq R_{min} \qquad (8.1.2.4.2)$$

$$\sum_r q_{pr} = frac_p \qquad \forall\, p \qquad (8.1.2.4.3)$$

$$q_{pr} \geq 0 \qquad \forall\, p, r \qquad (8.1.2.4.4)$$

In the objective function the operating costs per piece CO_{mpr} of part p produced on route r at machine m are multiplied by the fraction q_{pr} of total production flow which is produced via that route. Thus for a given routing vector $\underline{q}$ the average operating costs per piece is obtained. These costs are minimized over $\underline{q}$.

To ensure a minimum production rate R_{min} the production rate of the system $T_N(\underline{q})$ is set greater or equal to R_{min}. In constraint set (8.1.2.4.3) the required product rate $frac_p$ of

---

172   see chapter 4.2.2 and 7.5

part type p, i.e. the fraction of part p of total production, is set equal to the sum of fractions q_{pr} over all routes. The last constraint set (8.1.2.4.4) forces the decision variables q_{pr} to be positive or equal to zero.

Procedure for solution

Due to the additional constraint for a minimum production rate R_{min} a solution procedure based on the adapted form of the FD algorithm can not be applied. Thus the solution of the above model will be generated by an augmented Lagrangian function.[173] The function has the following form:

$$LF_a = \sum_m \sum_p \sum_r CO_{mpr} \cdot q_{pr} + \sum_p [\alpha_{1p} \cdot (\sum_r q_{pr} - frac_p) + w_1 \cdot (\sum_r q_{pr} - frac_p)^2]$$

$$+ \left[\begin{array}{ll} + \alpha_2 \cdot (R_{min} - T_N(\underline{q})) + w_2 \cdot (R_{min} - T_N(\underline{q}))^2 & \text{if } (R_{min} - T_N(\underline{q})) > -\alpha_2/2 \cdot w_2 \\ - \alpha_2^2/4 \cdot w_2 & \text{if } (R_{min} - T_N(\underline{q})) \leq -\alpha_2/2 \cdot w_2 \end{array} \right.$$

The throughput $T_N(\underline{q})$ of a given flow vector is evaluated by the throughput formula of closed queueing networks. To have convexity of model ROLP-NEW-(2) concavity of the throughput function must be ensured. Even though Appendix A proves that the delay function - which is reciprocally related to the throughput function - is convex, concavity of the throughput function can not be assumed a priori. In fact nonconcavity in the single-server case with more than two pallets in the system can be proved.[174] However, even if the model formulation above is not convex owing to the penalty functions in the augmented Lagrangian, the duality gap can be removed and the solution of the saddle point of LF_a is then an optimal solution for ROLP-NEW-(2).[175] To ensure that this optimal solution is not only local but global, quasiconcavity for the throughput function must be given.[176] Here Stecke has proven quasiconcavity in a single-server case and presumed, based on empirical evidence, that it also holds for the multiple-server case.[177]

Example

To demonstrate the algorithm an example is presented. For the part data refer to tab.8.1 in chapter 8.1.2.3. The operating costs for each route are given in tab.8.5.

---

173   for a description of this solution procedure refer to chapter 6.1.1.1

174   Stecke, K.E.: On the Nonconcavity of Throughput in Certain Closed Queueing Networks, in: Performance Evaluation, 6(1986), pp.296-297

175   see chapter 6.1.1.1

176   Bazaraa, M.S., Shetty, C.M.: Nonlinear Programming: Theory and Algorithms, New York 1979, p.148

177   see Stecke, K.E., Solberg, J.J.: The Optimal Planning of Computerized Manufacturing Systems, School of Industrial Engineering, Perdue University, Report No.20, West Lafayette, Indiana 1981, p.125

part	route	L/U St.	mill-1	mill-2	mill-3	vtl-1	vtl-2	transp.
part1	route1	20.00	10.00	15.00	-	-	-	-
	route2	30.00	10.00	-	-	10.00	-	-
part2	route1	15.00	-	-	10.00	5.00	10.00	-
	route2	10.00	10.00	-	-	15.00	-	-
part3	route1	30.00	5.00	3.00	3.00	10.00	-	-
	route2	30.00	-	5.00	10.00	-	4.00	-
	route3	20.00	-	20.00	20.00	-	-	-

tab. 8.5: Operating costs at each machine for each route

Again the same configuration of the flexible manufacturing system (tab.8.2) and the same product rates $frac_p$ (tab.8.3) as in chapter 8.1.2.3 are used. There are 20 pallets in the system and the required production rate R_{min} is 25,000 parts per period.

As a result of the above described algorithm the flow fractions of each route and the production rate are obtained (see tab.8.6).

	part1		part2		part3		
	route1	route2	route1	route2	route1	route2	route3
q_{pr}	0.285	0.004	0.022	0.373	0.131	0.184	0.000
T_{pr}	7114	101	546	9328	3283	4601	0

tab. 8.6: Flow fraction on each route

In fig.8.2 a comparison of throughput maximization to cost minimization is done. The fractions q_{pr} for all parts on all routes are shown for both models, as well as the resulting production rates.

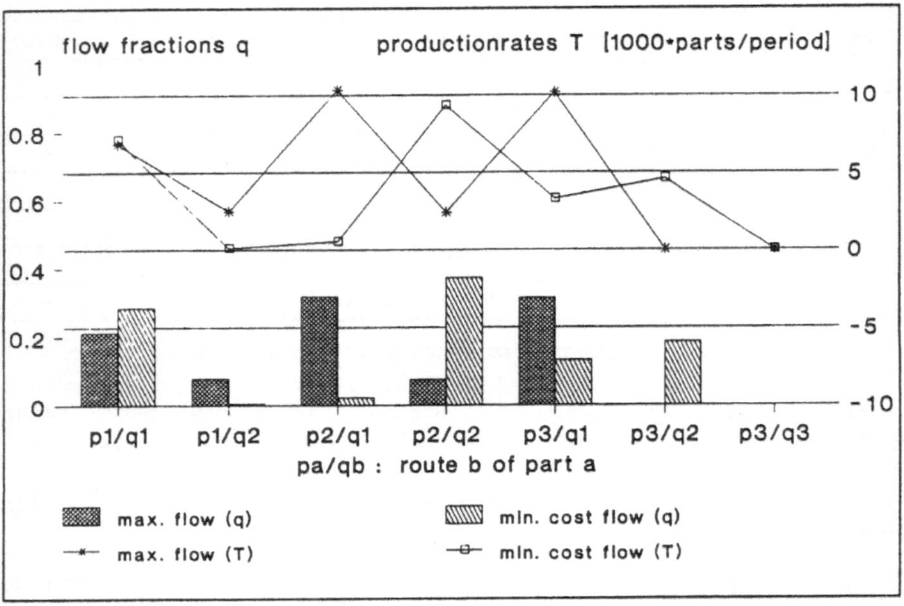

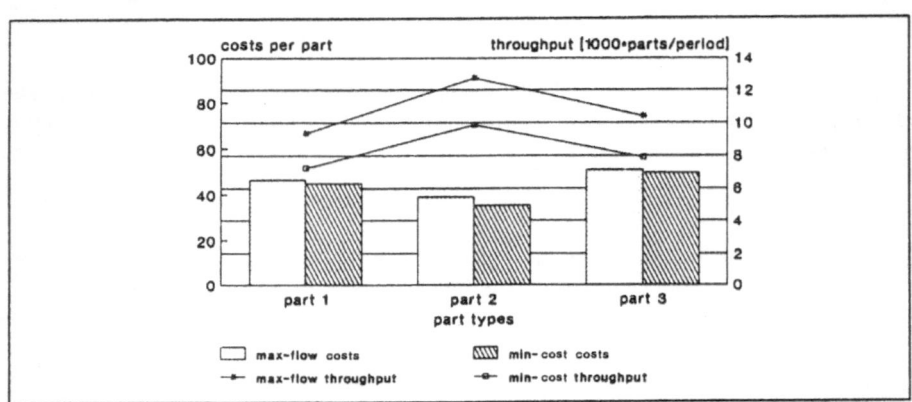

fig. 8.2: Comparison of routing fractions for max flow and min-cost flow

The different costs per part for throughput maximization and cost minimization are given in fig.8.3.

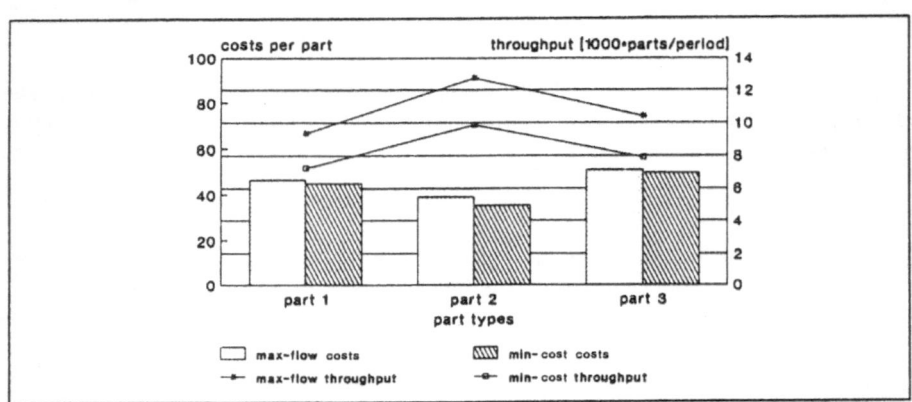

fig. 8.3: Comparison of costs per part for max flow and min-cost flow

Discussion

Input:
- Different alternative routes and their costs must be provided.
- Proper estimates for the Lagrangian multiplier and penalty weights must be provided at the beginning of the algorithm. For the Lagrangian multipliers a_{1p} the use of a value which reflects the average operating costs for the given part type p on a single route is suggested. The penalty weights must be chosen high enough to prevent a constraint violation. However excessive penalty weights at the beginning of the algorithm cause poor convergence rates.

Model:
- Dynamic system behavior modelling allows the application of this model to flexible machining systems.
- Possible cost reductions through the performance of a certain combination of operations are neglected.
- The model is limited to one universal pallet type. A system with several specialized pallet types yielding different system behavior is not considered.
- Further limitations are given by the stochastic assumptions of closed queueing network theory.[178]

Procedure:
- Assuming quasiconcavity of the throughput function, the global optimum is obtained.
- Compared with the adapted FDC algorithm, at each optimization step for q a multi-dimensional search technique can be applied. The convergence performance depends on the chosen initial values for the Lagrangian multipliers, the weight factors and q.

8.2 Models for capacity optimization

Models for capacity optimization have the quality that the types of equipment to be integrated into the flexible manufacturing system are already chosen, but the number of each type, i.e. the capacity of each type, has to be investigated. Thus the part flow through the system is given, but the required number of, for example, machines, vehicles, or pallets to fulfill given requirements must still be determined.

8.2.1 Optimal server and pallet capacity by Vinod and Solberg

The model by Vinod and Solberg allows the determination of the optimal number of machines, transport carriers and pallets to be incorporated into a flexible manufacturing system. The process plans of the parts to be produced on the system are fixed and no alternative plans are given. The optimization process itself is conducted by cost

---

178   see chapter 6.1.2.2

considerations, i.e. investment and operating costs.[179]

Model CALP-Vinod/Solberg

Index:
m   :   cell index

Decision variables:
N   :   number of pallets in the system
s_m :   number of machines of machine type m or number of carriers of the transportation system

Parameters:
C_m :   operating and capital investment costs per server (machine, load/unload station, transportation vehicle etc.) of type m
C_N :   operating and capital investment costs per pallet
R_{min} :   desired production rate
$T_M(s,N)$:   throughput of the system with M stations (cells and transportation system)
SW_O :   system workload
W_m :   workload at cell m

Objective function:

$$min \quad Z(\underline{s},N) = \sum_{m=1}^{M} (C_m \cdot s_m) + C_N \cdot N \tag{8.2.1.1}$$

Constraints:

$$T_M(\underline{s},N) \geq R_{min} \tag{8.2.1.2}$$

$$s_m \in Z^+ \ \forall \ m, \quad N \in Z^+ \tag{8.2.1.3}$$

In the objective function the factor C_m describes the operating and capital investment costs per server (CNC-machine, vehicle, load/unload station etc.), whereas C_N describes the operating and capital investment costs for each pallet (job). Multiplied by the number of servers s_m or by the number of pallets N respectively, the objective function represents total operating and capital investment costs. It is minimized under the constraint that the system must fulfill a desired production rate R_{min}. The actual throughput is evaluated by the throughput function $T_M(\underline{s},N)$ derived from closed queueing network theory.[180]

Procedure for solution

Below two solution procedures are described. First the original procedure proposed by Vinod and Solberg[181] is presented. Afterwards an alternative approach proposed by Dallery and Frein[182] is shown.

179   Vinod, B., Solberg, J.J.: The optimal design of flexible manufacturing systems, in: IJPR, 23(1985)23, pp.1141-1151
180   see chapter 6.1.1.2
181   Vinod, B., Solberg, J.J.: The optimal design of flexible manufacturing systems, in: IJPR, 23(1985)23, pp.1146-1150
182   Dallery, Y., Frein, Y.: An efficient Method to determine the optimal Configuration of a Flexible

Algorithm of Vinod and Solberg:

The solution procedure of the above model is complicated by the fact that the throughput function of the constraint is nonlinear and has to be evaluated in a recursive manner. That is why Vinod and Solberg have chosen an implicit enumerative algorithm, which is based on the following assumption of monotonicity for the throughput function of the classical closed queueing network:

Given $(s_1,\ldots,s_m,\ldots,s_M) \leq (s_1,\ldots,s_m+1,\ldots,s_M)$

then $T(s_1,\ldots,s_m,\ldots,s_M,N) \leq T(s_1,\ldots,s_m+1,\ldots,s_M,N)$ for $m=1,\ldots,M$

Thus, if another server (CNC-machine, vehicle, set-up table) is added to one of the cells (or the transportation system), throughput does not decrease.[183]

Similarly, for the number of pallets, the following monotonocity property can be used:[184]

$$T(N-1) \leq T(N)$$

The whole solution procedure, described below, consists of three steps.

First an upper bound $T_M{}^u$ and a lower bound $T_M{}^l$ of the throughput is evaluated using a given upper bound vector of the decision variables s_m (number of servers) and N (number of pallets). The variable W_m is the workload at each cell or the transportation system, obtained from the given process plans. After calculating an upper bound $T_M{}^u$ and a lower bound $T_M{}^l$ an estimate for the production rate R_{min} can be specified.[185]

$$T_M{}^u = \min_m \left[\frac{\min (s_m,N)}{W_m} \right] \qquad (8.2.1.4)$$

$$T_M{}^l = \left[\sum_{m=1}^{M} \frac{W_m}{\min (s_m,N)} \right]^{-1}$$

The next step consists of a bisection search procedure (fig.8.4). With the help of this heuristic a good starting solution $\underline{v}$ for the server variable $\underline{s}$ is obtained for reducing the number of evaluations in the following enumerative algorithm. The search procedure consists of two parts. In the first part a given start vector of the configuration $\underline{u}$ (with $\underline{u}$ greater then the yet unknown best solution $\underline{x}^M$) is bisected as long as its upper bound $T_M{}^u$ is larger than the given production rate R_{min}. If this is no more the case, an arithmetic mean of the largest nonfeasible configuration vector $\underline{y}$ and the smallest feasible vector $\underline{v}$ is calculated in the second part and the obtained values rounded to the next larger integer. This is continued until the two vectors $\underline{y}$ and $\underline{v}$ are identical. The vector $\underline{v}$ (or $\underline{y}$) is now the solution of the heuristic. The quality of this heuristic solution obtained depends largely on the as yet disregarded objective function and on the initial, arbitrarily chosen upper bound vector $\underline{u}$.

Manufacturing System, in: Proc. 2nd ORSA/TIMS Conf. on Flexible Manufacturing Systems: Operations Research Models and Applications, Ed.: K.E. Stecke and R. Suri, Amsterdam 1986, pp.272-275

183   A proof is given in Shanthikumar, J.G., Yao, D.D.: Optimal Server Allocation in a System of Multi-Server Stations, in: MS, 33(1987)9, pp.1173-1180

184   A proof is given in Suri, R.: A Concept of Monotonicity and Its Characterization for Closed Queueing Networks, in: OR, 33(1985)3, pp.606-624

185   Vinod, B., Solberg, J.J.: The optimal design of flexible manufacturing systems, in: IJPR, 23(1985)23, p.1144

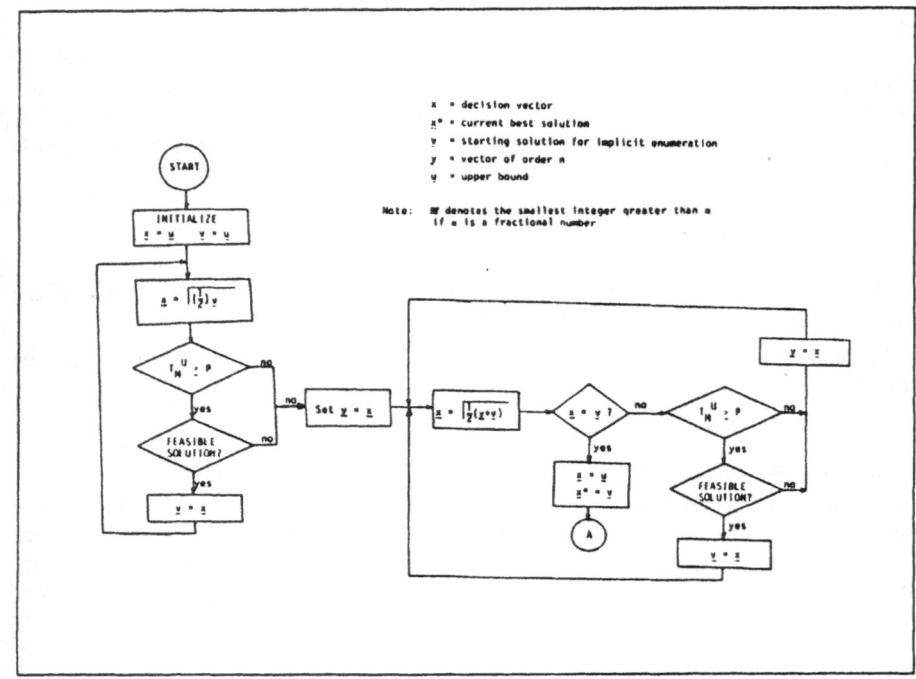

fig. 8.4: Bisection search procedure[186]

In the last step an implicit enumeration is performed between starting vector $\underline{v}$ obtained from the preceding heuristic and a given lower bound vector $\underline{l}$. The latter can be obtained by reformulation of (8.2.1.4). Bounds for each configuration vector are provided by the so far best solution Z^* of the objective function (at the beginning provided by the heuristic solution) and the estimation of the upper bound of the throughput T_M^u [see (8.2.1.1)] in comparison to the necessary production rate R_{min}.

Algorithm of Dallery and Frein

Dallery and Frein criticize the procedure of Vinod and Solberg as having the following drawbacks:[187]

-   the required initial solution is highly dependent on the choice of an arbitrary starting point;
-   the computational time rapidly increases with the number of stations, and the procedure is often infeasible.

Based on the assumption of an increasing cost function:

---

186   Vinod, B., Solberg, J.J.: The optimal design of flexible manufacturing systems, in: IJPR, 23(1985)23, p.1147
187   Dallery, Y., Frein, Y.: An efficient Method to determine the optimal Configuration of a Flexible Manufacturing System, in: Proc. 2nd ORSA/TIMS Conf. on Flexible Manufacturing Systems: Operations Research Models and Applications, Ed.: K.E. Stecke and R. Suri, Amsterdam 1986, p.270

Given $(s_1, \ldots, s_m, \ldots, s_M, N) \leq (s_1, \ldots, s_m+1, \ldots, s_M, N)$

then $Z(s_1, \ldots, s_m, \ldots, s_M, N) \leq Z(s_1, \ldots, s_m+1, \ldots, s_M, N)$ for $m=1, \ldots, M$

they suggest an alternative solution procedure, which also incorporates three steps.

In the first step of the solution procedure a lower bound configuration vector $(\underline{s}^b, N^b)$ is evaluated by ABA bounds (Asymptotic Bound Analysis).[188]

$$s_m \geq R_{min} \cdot W_m$$

The number of servers s_m (CNC-machines, vehicles, set-up tables) at each station m (cell or transportation system) is evaluated through the required production rate R_{min} multiplied by the workload W_m at that station. The number of pallets is estimated by the product of R_{min} with the system's workload SW_0.

$$N \geq R_{min} \cdot SW_0$$

In the next step with the help of the lower bound configuration vector $(\underline{s}^b, N^b)$ a heuristic search procedure tries to improve this vector and to find an optimal or a near optimal solution.

This is done according to a marginal allocation scheme. At each step the station (cell, transportation system) that provides the steepest ascent δX_m of increase in the production rate per increase on overall costs is enlarged:

$$\delta X_m = \begin{cases} \dfrac{T(\underline{s} + \underline{1}_m, N^b) - T(\underline{s}, N^b)}{Z(\underline{s} + \underline{1}_m, N^b) - Z(\underline{s}, N^b)} & \text{if } Z(\underline{s} + \underline{1}_m, N^b) < Z^h \\ 0 & \text{if } Z(\underline{s} + \underline{1}_m, N^b) \geq Z^h \end{cases}$$

with $\underline{1}_m$ being a zero vector having the value 1 at position m

After the server is added, at each step the number of pallets N is increased until either the costs are higher than the costs of the current best solution or the required production rate R_{min} is reached. If the latter is the case, the new solution yields the so far best solution Z^*. The procedure ends if $\delta X_m=0$ for all m. The best solution found yields the heuristic solution Z^h with the configuration vector $(\underline{s}^h, N^h)$.

The configuration vector $(\underline{s}^h, N^h)$ and its heuristic solution Z^h of the objective function form the basis of the last step, which consists of an implicit enumeration procedure. At the beginning of the procedure a current best solution is provided by the solution of the heuristic. Then, starting from the lower bound configuration, all possible configurations are systematically enumerated. At each step, the cost Z of the current configuration $(\underline{s}, N)$ is compared to the cost Z^* of the current best solution $(\underline{s}^*, N^*)$ and, if the costs of the current solution are less, the constraint of minimal production is checked. If the constraint is not violated a new best solution is obtained. If however calculated costs are higher, owing to a monotone increasing objective function (see above) no larger configuration vector $(\underline{s}', N')$ with $(\underline{s}, N) < (\underline{s}', N')$ has to be considered.

---

188   Denning, P.J., Buzen, J.P.: The operational analysis of queueing network models, in: Computing Surveys, 10(1978)3, pp.225-262; Kleinrock, L. Queueing Systems, Vol.2 New York 1976 pp.219-225

Discussion

Input:
-   Vinod and Solberg do not further specify the nature of capital investment and operating costs per unit of a server or a job. If capital investment costs are analyzed in conjunction with operating costs, attention must be given to the fact that operating costs are directly output dependent, whereas investment costs are only output dependent by stages.

Model:
-   Dynamic system behavior modelling allows this model to be applied to flexible machining systems.
-   The model is limited to one universal pallet type. A system with several specialized pallet types yielding different system behavior is not considered.
-   The model only considers constant demand. This implies that changing levels in production requirements during the existence of such a flexible manufacturing system are not anticipated.
-   Further limitations are given by the stochastic assumptions of closed queueing network theory.[189]

Procedure:
-   An optimal solution is obtained.

8.2.2 Optimal server capacity with specialized pallets by Dallery and Frein

In this chapter an extension on the model CALP-Vinod/Solberg of Vinod and Solberg (see chapter 8.2.1) is given, considering systems with specialized pallet types.[190] The objective of the model is to find the optimal capacity of the cells and of the transportation system, so that costs assignable to each server are minimized. This is done under the constraint of the fulfillment of a given minimum production rate for each pallet type. The part family which has to be produced, and its process plans are given. The decision variable, i.e. the capacity of each cell or the transportation system, is described by the number s_m of identical, parallel servers (machines, vehicles etc.). The number of pallets for each pallet type c is given.

Model CALP-Dallery/Frein

Index:
m   :   cell index

---

189   see chapter 6.1.2.2
190   Dallery, Y., Frein, Y.: An Efficient Method to determine the Optimal Configuration of a Flexible Manufacturing System, in: Annals or OR, 15(1988), pp.207-225

Decision variable:

s_m :   number of parallel servers at cell m, with i = 1,..,M

Parameters:

$Z(\underline{s})$:   increasing cost function for a given configuration, dependent on the number of servers s_m

R_{minc} :   minimal required production rate for pallet type c

$T_c(\underline{s})$:   throughput for pallet type c

Objective function:

$$\min \; Z(\underline{s}) \tag{8.2.2.1}$$

Constraints:

$$T_c(\underline{s}) \;\geq\; R_{minc} \qquad \forall \; c \tag{8.2.2.2}$$

$$s_m \in Z^+ \qquad \forall \; m \tag{8.2.2.3}$$

In the objective function $Z(\underline{s})$ the server costs are minimized. Hereby $Z(\underline{s})$ is an increasing function, i.e. if $\underline{s} \geq \underline{s}'$ then $Z(\underline{s}) \geq Z(\underline{s}')$. This is achieved under the constraint of a required production rate R_{minc} for each pallet type c. The actual throughput $T_c(\underline{s})$ of each part type is obtained through mean value analysis.[191] Note that the number of pallets of each type is given.

Procedure for solution

To solve the above model a three step procedure is given. First lower bounds for the number of servers at each cell/transportation system s_m^b are derived from the forced flow law and utilization law:[192]

$$s_m^b \;\geq\; U_m \;=\; \sum_{c=1}^{C} U_{mc} \;=\; \sum_{c=1}^{C} W_{mc} \cdot R_{minc} \qquad \forall \; m$$

Here the utilization at each cell m must be less than or equal to the number of servers at that cell. The utilization of cell m is given by the summation over all pallet types c of the workloads W_{mc} multiplied by the minimal production rate R_{minc}. Next, with the help of a marginal allocation approach, a heuristic solution is generated. The procedure can be stated as follows:

Heuristic for capacity optimization with specialized pallets
Step 0:   set $s_m = s_m^b$ $\forall$ m

---

191   for a description of mean value analysis see chapter 6.1.2.3.

192   Denning, P.J., Buzen, J.P.: The operational analysis of queueing network models, in: Computing Surveys, 10(1978)3, p.225

Step 1:

$$\text{set } \delta_{mc} = \begin{cases} T_c(\underline{s}+\underline{1}_m) - T_c(\underline{s}) & \text{if} \quad T_c(\underline{s}) < R_{minc} \\ & \text{or} \quad T_c(\underline{s}+\underline{1}_m) < R_{minc} \\ 0 & \text{if} \quad T_c(\underline{s}) \geq R_{minc} \\ & \text{and} \quad T_c(\underline{s}+\underline{1}_m) \geq R_{minc} \end{cases}$$

with $\underline{1}_m$ being a zero vector having the value 1 at position m

$$\text{calculate } \Delta F_m = \frac{\displaystyle\sum_{c=1}^{C} \delta_{mc}}{Z(\underline{s}+\underline{1}_m) - Z(\underline{s})} \qquad \forall \; m$$

if $\max_m \Delta F_m = 0$ stop

set m^* as the index of: $\max_m \Delta F_m$

Step 2:

add a piece of equipment of type m^* to $\underline{s}$: $\quad \underline{s} = \underline{s} + \underline{1}_{m^*}$

update $Z^h(\underline{s}) = Z(\underline{s})$

go to step 1.

alg. 8.2: Greedy heuristic for capacity optimization with specialized pallets

Finally, in the third step an implicit enumeration is carried out to find the global optimum. With the help of the heuristic solution Z^h as an upper bound and a lower bound given by the configuration vector $\underline{s}^b$ from the first step, the enumeration is accelerated.

Discussion

Input:
- The cost function for server costs is not specified.
- The number of pallets of each type must be provided as input data.

Model:
- Dynamic system behavior modelling allows this model to be applied on flexible machining systems.
- The model considers specialized pallet types.
- The model only considers constant demand. This implies that changing levels in production requirements during the existence of such a flexible manufacturing system are not anticipated.
- Further limitations are given by the stochastic assumptions of closed queueing network

theory.[193]

Procedure:
- The optimal solution can be obtained.

8.2.3 Optimal allocation of identical servers by Yao and Shanthikumar

This model by Shanthikumar and Yao allocates a given number of identical machines between M cells of a flexible manufacturing system. It is assumed that these machines are versatile enough to perform all operations required in the system. However, if a machine is assigned to a cell, limited tool capacity allows this machine to perform only those operations assigned to that cell. Thus it looses its versatile character. The objective of the allocation is to maximize total throughput for a given workload distribution over the cells, i.e. fixed process plans for the parts are given.[194]

Model CALP-Yao/Shanthikumar-(1)

Index:
m   :   cell index

Decision variable:
s_m :   number of parallel servers at cell m

Parameters:
S   :   number of available identical machines
W_m :   workload at cell m
M   :   number of cells
N   :   number of pallets
$T(\underline{s})$:   system throughput

Objective function:
$$\max_{\underline{s} \in X} \quad T(\underline{s}) \tag{8.2.3.1}$$

$$\text{where } X = \{\underline{s}: s_m \epsilon Z^+, m=1,\ldots,M; \sum_{m=1}^{M} s_m = S\} \tag{8.2.3.2}$$

$$\text{and } T(\underline{s},\underline{W}) = G(N-1)/G(N) \tag{8.2.3.3}$$

In the objective function the throughput $T(\underline{s})$ of the system is maximized. Thereby its evaluation is done by closed queueing network theory with the help of the normalization constant G having N, respectively N-1 pallets in the system and a workload distribution over all cells of $\underline{w}$ (see chapter 6.1.2.2). Maximization is done over set X given by all $\underline{s}$ for

---

193   see chapter 6.1.2.2
194   Shanthikumar, J.G., Yao, D.D.: On server allocation in multiple center manufacturing systems, in: OR, 36(1988)2, pp.333-342; Yao, D.D., Shanthikumar, J.G.: Some resource allocation problems in multi-cell systems, in: Proc. 2nd ORSA/TIMS Conf. on Flexible Manufacturing Systems: Operations Research Models and Applications, edited by K.E. Stecke and R. Suri, Amsterdam 1986, pp.249-250

which the number of servers (machines) s_m at each cell m is a positive integer and the sum over all servers (machines) is equal to S.

Procedure for solution

Shanthikumar and Yao propose two algorithms - an exact procedure and a greedy heuristic - to solve the above model. For the exact procedure they prove the following theorem:

Theorem 5 (decreasing property of the optimal allocation): An optimal allocation, say $\underline{s}^*$, to the above model satisfies $s_1^* \geq s_2^* \ldots \geq s_M^*$ when the workloads are ordered in the following form: $W_1 \geq W_2 \ldots \geq W_M$.[195]

Then the set of allocations D for which this decreasing property holds is generated by recursion:

For two cells the set of allocations $D(n,2)$ for n machines is

$$D(n,2) = \{(n,0),(n-1,1),\ldots,(n-[n/2],[n/2])\}$$

For m + 1 cells we have

$$D(n,m+1) = \{(\underline{o}+k\underline{1},k) \mid \underline{o} \in D(n-(m+1)k,m), k=0,1,\ldots,[n/(m+1)]\}$$

where $\underline{1}$ is an m-dimensional vector with all components being one.

Based on these results, they suggest a three step search procedure. In the first step all possible allocations D for which the decreasing property holds, are generated by the above recursive formula. Furthermore, upper bounds are calculated for each configuration vector. In the next step throughput is calculated for the configuration vector $\underline{s}^1$ with the largest upper bound via closed queueing network theory and all other configurations which have an upper bound less then the just calculated throughput $TH(\underline{s}^k)$ are eliminated. The last step is repeated with $\underline{s}^2 \ldots \underline{s}^k$, until all configuration vectors are eliminated. The optimal configuration vector $\underline{s}^*$ is given by the vector for which the largest throughput was calculated.

Solution procedure for CALP-Yao/Shanthikumar-(1)

Step 1:

generate the set $D(0,M)$

for each $\underline{o} \in D(0,M)$, set $\underline{s}=\underline{o}+\underline{1}$ and calculate

the balanced load bound $UB(\underline{s})$ according to Eager and Sevcik[196]:

$$TH(\underline{s},\underline{W}) \leq TH(1,\underline{W}/\underline{s}) \leq \frac{N \cdot M}{(N+M-1) \cdot \sum_{m=1}^{M} (W_m/s_m)} = UB(\underline{s})$$

---

195   A proof can be found in Shanthikumar, J.G., Yao, D.D.: On server allocation in multiple center manufacturing systems, in: OR, 36(1988)2, pp.335-336
196   Eager, D.L., Sevcik, K.C.: Performance Bound Hierarchies for Queueing Networks, in: ACM Trans.

```
Step 2:

let s¹ = arg max UB(s) (i.e. UB(s¹) is the largest upper bound)

compute TH(s¹)

eliminate all s (if any) that satisfy UB(s)≤TH(s¹)
```

```
Step 3:

for the remaining o ∈ D(O,M), repeat Step 2 to generate s²,...,sᴷ,

until all o ∈ D(O,M) are eliminated

stop: solution is given by s* = arg max [TH(sⁱ), i=1,...K].
```

alg. 8.3: Optimal server allocation

The greedy heuristic is performed by Fox's marginal allocation scheme:[197]

Suppose there are S extra machines to be allocated. These extra machines are then allocated one at a time in accordance with the following rule: assign the machine to the work station (cell) that will generate the largest increase in throughput. If this is the case for more than one machine, then any one of the alternative allocations that give the maximum increase can be considered.

Discussion

Model:

- Shanthikumar and Yao consider in this model only identical machines. This assumption seems to be questionable. In general flexible manufacturing systems contain different machines types which can only perform certain kinds of operations, for example load/unload operations by load/unload stations, washing by washing machines, milling by milling machines etc..

- In the presentation the transportation system is completely neglected. If it is introduced into the model with a fixed number of vehicles s and a given workload W, the decreasing property theorem and therefore the exact solution procedure is in general no longer applicable.

- Dynamic system behavior modelling allows this model to be applied to flexible machining systems.

- The model is limited to one universal pallet type. A system with several specialized pallet types yielding a different system behavior is not considered.

- It is assumed that there is no restriction concerning the demand for finished parts. This is for example the case when the flexible manufacturing system is the bottleneck of the

Comp. Syst., 1(1983), pp.99-115

197   Shanthikumar, J.G., Yao, D.D.: On server allocation in multiple center manufacturing systems, in: OR, 36(1988)2, p.338; Fox, B.: Discrete Optimization via Marginal Analysis, in: MS, 13(1966), pp.210-216

production process.

- Further limitations are given by the stochastic assumptions of closed queueing network theory.[198]

Procedure:
- An optimal solution is obtained.

8.2.4 Optimal allocation of nonidentical servers by Yao and Shanthikumar

In the previous model by Yao and Shanthikumar identical servers were allocated to the cells in a flexible manufacturing system. This model extends the previous one by allocating servers to different cells, which are not necessarily identical. The objective is, to maximize the difference between increasing concave profit functions for the throughput of each cell and convex cost functions for the number of servers (machines, load/unload stations etc.) allocated to each cell.

Model CALP-Yao/Shanthikumar-(2)

Index:
m :   cell index

Decision variables:
s_m :   number of parallel servers at cell m

Parameters:
S :   number of available identical machines
$\lambda_m(k)$:   arrival rate to cell m, given the number of jobs there is k, $0 < \lambda_m(k) < \infty$ if $k < N_m$, and $\lambda_m(k) = 0$ if $k \geq N_m$
$T_m(s_m)$:   throughput of station m, given that s_m servers have been allocated to it
$f_m(T_m)$:   profit at cell m as a function of its throughput T_m
$g_m(s_m)$:   cost at cell m as a function of the number of servers s_m allocated to it
M :   number of cells
μ_m :   the service rate per server at cell m
N_m :   buffer limit at cell m

Objective function:

$$\max_{\underline{s} \in S} \sum_{m=1}^{M} f_m(T_m) - g_m(s_m) \tag{8.2.4.1}$$

$$\text{where } S \equiv \{\underline{s}: s_m \in Z^+ \text{ and } s_m \leq N_m \text{ for } m=1,\ldots,M ; \sum_{m=1}^{M} s_m \leq S\} \tag{8.2.4.2}$$

In the objective function the sum of the differences between the profit function $f_m(T_m)$ of each cell and its cost function $g_m(s_m)$ are maximized. The profit of each cell depends on

---

198   see chapter 6.1.2.2

its throughput T_m. The latter is evaluated by a birth death queue:[199]

$$T_m(s_m) = \sum_{k=0}^{N_m-1} \lambda_m(k) \cdot w_m(k) \bigg/ \sum_{k=0}^{N_m} w_m(k)$$

$$\text{with } w_m(0)=1 \text{ and } w_m(k) = \prod_{l=1}^{k} \lambda_m(l-1)/[\mu_m \cdot \min(l,s_m)] \text{ for } k\geq 1$$

The maximization is done subject to the constraints that the number of servers (machines) at each cell is a positive integer, so that the number of servers s_m does not exceed the available buffer space N_m at each cell m, and the number of servers (machines, load/unload stations, transportation vehicle etc.) in the system is less than or equal to S.

Procedure for solution

A greedy algorithm is performed, based on the marginal allocation scheme of Fox.[200] It finds the optimal solution, as the profit function $f_m(T_m)$ and the throughput $T_m(s_m)$ is increasing concave in s_m and the cost functions are convex. The procedure is as follows:[201]

The servers are allocated one at a time in accordance with the rule: assign the server to the work station (cell) that will generate the largest increase in the objective function. Once the number of servers at a cell m reaches its buffer limit N_m or the increase $F_m(s_m+1)-F_m(s_m)$, with $F_m=f_m(T_m)-g_m(s_m)$ becomes nonpositive, eliminate that cell from further consideration. Continue this procedure until either all S servers have been allocated or all cells have been eliminated, whichever comes first.

Discussion

Input:
- Proper estimates for the arrival rate for each cell must be available.
- The authors give no specification for the form and structure of the profit- and cost functions. However, the assignment of a reasonable profit function, depending on the throughput of the cell, seems to be crucial.

Model:
- Dynamic system behavior modelling allows the application of this model to flexible machining systems. However, the modelling of each station as a single queue, and not as a network of queues, neglects dependencies of the part flow between the cells.
- It is assumed that there is no restriction concerning the demand for finished parts. This is for example the case when the flexible manufacturing system is the bottleneck of the production process.
- The model is limited to one universal pallet type. A system with several specialized

199   Shanthikumar, J.G., Yao, D.D.: Optimal server allocation in a system of multi-server stations, in: MS, 33(1987)9, p.1174
200   Fox, B.: Discrete Optimization via Marginal Analysis, in: MS, 13(1966), pp.210-216
201   Shanthikumar, J.G., Yao, D.D.: Optimal server allocation in a system of multi-server stations, in: MS, 33(1987)9, p.1176

pallet types yielding different system behavior is not considered.

Procedure:
-   An optimal solution is obtained.

8.2.5 Optimal buffer allocation by Yao and Shanthikumar

This model for optimal buffer allocation considers a flexible manufacturing system, where each cell can comprise different machine types.[202] The authors assume that there is only limited space available for the system and thus the maximum number of parts in the whole system is restricted to a given number N.[203] The problem is to allocate these parts between the cells, so that costs are minimized. The number of parts which each cell can simultaneously contain is considered to be equal to the number of necessary buffer spaces for each cell.

Model CALP-Yao/Shanthikumar-(3)

Index:
m :   cell index

Decision variables:
N_m :   number of buffers in cell m

Parameters:
$f_m(T_m(N_m))$:   profit function for cell m with throughput T_m and N_m buffers
$g_m(N_m)$:   cost function for the allocation of N_m buffers
N :   number of available buffers, i.e. pallets in the system

Objective function

$$\max \quad F(\underline{N}) = \sum_{m=1}^{M} f_m(T_m(N_m)) - g_m(N_m) \tag{8.2.5.1}$$

Constraint

$$\sum_{m=1}^{M} N_m \leq N \qquad N_m \in Z^+ \tag{8.2.5.2}$$

In the objective function for each cell the difference between the profit function $f_m(T_m(N_m))$, dependent on throughput and therefore also on the number of buffers in the cell, and the cost function $g_m(N_m)$ for the allocation of N_m buffers in cell m is evaluated. Then the sum over all differences is maximized.

---

202   see definitions in chapter 2.1
203   Yao, D.D., Shanthikumar, J.G.: Some resource allocation problems in multi-cell systems, in: Proc. 2nd ORSA/TIMS Conf. on Flexible Manufacturing Systems: Operations Research Models and Applications, edited by K.E. Stecke and R. Suri, Amsterdam 1986, pp.245-249; Shanthikumar, J.G., Yao, D.D.: Optimal Buffer Allocation in a Multicell System, in: The International Journal of Flexible Manufacturing Systems, 1(1989), pp.347-356

The constraint (8.2.5.2) ensures that the total number of buffers assigned to the system will be less than or equal to N, and that the number of buffers N_m for each cell is a positive integer.

Procedure for solution

Yao and Shanthikumar propose to use either the closed queueing network model of Gordon and Newell[204] or Jackson's open queueing network model[205] for throughput evaluation of each cell. They assume for all m = 1,..,M , that $f_m(T_m(N_m)$ is an increasing concave function, and $g_m(N_m)$ a convex function. Therefore in the above optimization problem the functions $F_m(N_m) = f_m(T_m(N_m)) - g_m(N_m)$ are concave in N_m for all m.

To find the optimal solution for the above problem, Fox's marginal allocation approach is applied.[206] This approach can be roughly described as follows:[207]

Allocate buffers one at a time. Every time assign a buffer to the cell which shows the largest increase $\Delta F_m(N_m) = F_m(N_m+1) - F_m(N_m)$. Once a cell's buffer size reaches the point where $\Delta F_m(N_m) \leq 0$, eliminate that cell from further consideration. Continue this procedure, until either all buffers are allocated or all cells are eliminated.[208]

Discussion

Model:

- The optimal buffer allocation with the help of the open or closed queueing network model as proposed by Shanthikumar and Yao seems questionable.

  The authors assume: "Suppose the dimension (space) of each cell is measured by the maximum total number of parts allowed to be simultaneously processed within the cell at any time. We shall refer to this number as the buffer size of a cell."[209]

  A basic assumption of both network models however is to have an *unlimited* number of buffers in front of every station or, if the pallet number is fixed to N_m in the system, to have at least as much buffer space in front of every station as there are pallets in the cell (i.e. the buffer size at each station in the cell must be at least as large as N_m).[210] Therefore the evaluation of system parameters with the help of standard open or closed queueing network models does not reflect the fact that the number of parts is identical with the number of buffers in the cell. The buffersize should be given as the product of

204   see chapter 6.1.2.2 and Gordon, W.J., Newell, G.F.: Closed Queueing Networks with Exponential Servers, in: OR, 15(1967), pp.252-267

205   Jackson, J.R.: Jobshop-Like Queueing Systems, in: MS, 10(1963), pp.131-142

206   Fox, B.: Discrete Optimization via Marginal Analysis, in: MS, 13(1966), pp.210-216

207   Yao, D.D., Shanthikumar, J.G.: Some resource allocation problems in multi-cell systems, in: Proc. 2nd ORSA/TIMS Conf. on Flexible Manufacturing Systems: Operations Research Models and Applications, edited by K.E. Stecke and R. Suri, Amsterdam 1986, p.248

208   Note that in the original model formulation by Shanthikumar and Yao constraint (8.2.5.2) is an equality. Above the constraint has been changed to an inequality. This is because it is possible that all cells have been eliminated before the actual number of buffers reaches N.

209   Yao, D.D., Shanthikumar, J.G.: Some resource allocation problems in multi-cell systems, in: Proc. 2nd ORSA/TIMS Conf. on Flexible Manufacturing Systems: Operations Research Models and Applications, edited by K.E. Stecke and R. Suri, Amsterdam 1986, p.247

210   Jackson, J.R.: Jobshop-Like Queueing Systems, in: MS, 10(1963), pp.131-142; Gordon, W.J., Newell, G.F.: Closed Queueing Networks with Exponential Servers, in: OR, 15(1967), pp.252-267

the number of parts (pallets) N_m multiplied by the number of stations M in the cell to ensure the condition of *unlimited* bufferspace at each station.

- For the form and structure of the profit and cost functions the authors give no specification. In particular, the assignment of a reasonable profit function, depending on the throughput of the cell, seems to be crucial.

- Dependencies of the part routing between the cells are neglected.

- It is assumed that there is no restriction concerning the demand for finished parts. This is for example the case when the flexible manufacturing system is the bottleneck of the production process.

- The model is limited to one universal pallet type. A system with several specialized pallet types yielding a different system behavior is not considered.

Procedure:
- An optimal solution is obtained.

8.2.6 Optimal pallet capacity for specialized pallet types by the algorithm

PANORAMA

The model formulation of PANORAMA (PAllet Number Optimization for Refined Automated MAnufacturing) describes the determination of the optimal number of pallets of each pallet type that should circulate in a flexible manufacturing system. Besides the pallet numbers, it considers a given configuration of the flexible manufacturing system, i.e. all other pieces of equipment have already been determined. The optimization is done under consideration of a given part mix and a maximum throughput time D_{max}.

Model CALP-Solot (PANORAMA)

Index:
c   :   pallet type index

Decision variable:
N_c :   number of pallets of type c in the system, $c = \{1,..,C\}$

Parameters:
ϵ :   return-on-investment correction factor
d_c :   pallet mix, i.e. the fraction that the throughput of pallet type c has of the system's throughput
D_{max} :   maximal throughput time
$D_c(N)$:   throughput time for pallet type r
$T(N)$:   throughput rate of the system
$T_c(N)$:   throughput rate of the system for pallet type c

Objective function:

$$\max_{\underline{N}} \frac{\min_{1 \leq c \leq R} \left[\dfrac{T_c(\underline{N})}{d_c} \right]}{(1+\epsilon)^N} \qquad (8.2.6.1)$$

Constraints:

$$D_c(\underline{N}) \leq D_{max} \quad \forall \, c \qquad (8.2.6.2)$$

$$\sum_c N_c = N \qquad (8.2.6.3)$$

$$N_c \in Z^+ \quad \forall \, c \qquad (8.2.6.4)$$

In the objective function first the value of the term:

$$\min_{1 \leq c \leq C} \left[\frac{T_c(\underline{N})}{d_c} \right] \qquad (8.2.6.5)$$

is evaluated. The external given factor d_c fixes the fraction that the throughput rate of pallet type c has of the throughput $T(\underline{N})$ of the flexible manufacturing system, i.e. $d_c = T_c(\underline{N})/T(\underline{N})$. The throughput rate $T_c(\underline{N})$ depends on the decision variables given by the vector $\underline{N}$, the number of pallets of each type c being in the system. Therefore the value of term (8.2.6.5) yields the maximal possible production rate of total production, if the flexible manufacturing system is the bottleneck. The following example given by Solot illustrates this:[211]

Consider a final product consisting of $k_1 = 3$ parts of type 1, $k_2 = 1$ part of type 2 and $k_3 = 1$ part of type 3. Only part types 1 and 2 are produced on the flexible manufacturing system. These parts require different pallet types. The pallet mixes are $d_1 = 0.75$ and $d_2 = 0.25$. Suppose for a given vector $\underline{N}$ of the pallet distribution the corresponding production rates are $T_1 = 6$ and $T_2 = 2.5$. In this case formula (8.2.6.5) yields 8 and the number of batches, i.e. the number of products that can be produced per unit of time, as 2. However, this is only correct under the assumption that part 3 is available in the necessary amount ($T_3 \geq 2$).

If the term (8.2.6.5) were maximized, the resulting number of pallets of each type would be infinite. To prevent this and to consider pallet costs, a return-on-investment correction factor is introduced. For every newly added pallet it consists of:

$$\frac{1}{(1+\epsilon)}.$$

Thus it is assumed that a company will purchase an extra pallet if this expense leads to a production increase greater than some threshold denoted by ϵ.[212] Consequently the term

---

211   Solot, P.: Optimizing a Flexible Manufacturing System with several Pallet Types, Working Paper O.R.W.P. 87/17, Ecole Polytechnique Fédéral de Lausanne, Département de Mathématiques, Lausanne Sept. 1987, p.7

212   Solot, P.: Optimizing a Flexible Manufacturing System with several Pallet Types, Working Paper O.R.W.P. 87/17, Ecole Polytechnique Fédéral de Lausanne, Département de Mathématiques, Lausanne Sept. 1987, p.7

(8.2.6.5) which stands for the production rate of total production, is divided by $(1 + \epsilon)^N$ and thus in some way normalized. This normalized production rate then forms the objective function.

In the first constraint set it will be ensured that the maximal throughput time D_{max} is not exceeded for any part. The remaining constraints restrict the number of pallets N_c of type c to positive integers, and the resulting sum over all pallet types to N.

Procedure for solution

For the solution of the above model Solot proposes a heuristic procedure which is based on the following two conjectures:[213]

Conjecture 4: The addition of one pallet of type c to the system leads to an increase in the production rate of pallet type c and a decrease in the production rates of the other pallet types.

Conjecture 5: The suppression of one pallet of type c from the system leads to a decrease in the production rate of pallet type c and an increase in the production rates of the other pallet types.

Based on these conjectures the following two conclusions can be drawn:

a) adding a pallet of type c^*, or

b) suppressing a pallet of another type,

can lead to an increase in the value of the objective function. Here c^* refers to the index which determines the minimum in term (8.2.6.5).

Thus the foundations of the heuristic are laid. Beginning at a pallet distribution vector of $\underline{N}^0 = \{1,..,1\}$ at each iteration the neighborhood of $\underline{N}$ is examined by adding a pallet of type c^* or suppressing single pallets which are of another type (i.e. $c \neq c^*$). The pallet distribution which yields the largest increase in the objective function while not violating the constraints is compared with the existing best solution. If it improves the best solution, the latter is updated. Then a new iteration step is performed, taking the best distribution found during the last step and examining its neighborhood. The algorithm ends if no improvement of the objective function can be made after a given finite number of iterations.

For the evaluation of the system parameters Solot refers to a queueing network program called MULTIQ.[214]

Discussion

Input:
- The selection of a reasonable value for the return-on-investment coefficient ϵ seems to

213   Solot, P.: Optimizing a Flexible Manufacturing System with several Pallet Types, Working Paper O.R.W.P. 87/17, Ecole Polytechnique Fédéral de Lausanne, Département de Mathématiques, Lausanne Sept. 1987, p.10

214   a description of the program MULTIQ can be found in: Solot, P., Bastos, J.M.: Choosing a Queueing Model for FMS, Working Paper O.R.W.P. 87/05, Ecole Polytechnique Fédéral de Lausanne, Département de Mathématiques, Lausanne 1987

be crucial.

Model:

- The model is limited to optimizing the number of each pallet type. However, other pieces of equipment play an important role in the design of a flexible manufacturing system and their effect is significant during optimization.

- Dynamic system behavior modelling allows this model to be applied to flexible machining systems.

- Further limitations are given by the stochastic assumptions of closed queueing network theory.[215]

Procedure:

- Only a heuristic solution is obtained.

8.2.7 A new model for server capacity optimization with budget constraint

In this chapter a new model is proposed which can be seen as a mixture between the model CALP-Vinod/Solberg by Vinod and Solberg (see chapter 8.2.1) and the server allocation models CALP-Yao/Shanthikumar-(1),(2) by Yao and Shanthikumar (see chapter 8.2.3 and 8.2.4). The objective of the model is to find the optimal capacity of the transportation system and of the cells, so that total throughput is maximized. This is done under the constraint of a limited budget BU. The part family, which has to be produced and the process plans of its part types are given as well. The decision variable, i.e. the capacity of each cell m and the transportation system, is described by the number s_m of identical parallel servers (machines, vehicles etc.). If the number of pallets is included in the decision process, the variable N refers to the number of pallets.

Model CALP-NEW-(1)

Index:
m   :   cell index
p   :   part type index

Decision variables:
s_m :   number of parallel servers at cell m or at the transportation system respectively, with m = 1,..,M
N   :   number of pallets in the system

Parameters:
BU   :   limited budget available
$g_m(s_m)$:   cost function for cell costs, dependent on the number of servers at cell m
$g_{M+1}(N)$:   cost function for pallet costs
$T(\underline{s})$:   system throughput, dependent on the number of servers at each station

---

215   see chapter 6.1.2.2

Objective function:

max $T(\underline{s}, N)$ (8.2.7.1)

Constraints:

$$\sum_{m=1}^{M} g_m(s_m) + g_{M+1}(N) \leq BU$$ (8.2.7.2)

$s_m \in Z^+$ $\forall\, m, \quad N \in Z^+$ (8.2.7.3)

In the objective function the throughput $T(\underline{s}, N)$ of the system is maximized. The evaluation of this function is done by closed queueing network theory using the normalization constant G having N or respectively N-1 pallets in the system (see chapter 6.1.2.2):

$T(\underline{s}, N) = G(N-1)/G(N)$

The first constraint (8.2.7.2) ensures that the summation over the cell costs $g_m(s_m)$ plus pallet costs $g_{M+1}(N)$ is less than the available budget BU. Note that the cost function $g_m(s_m)$ of each cell ($g_{M+1}(N)$ for pallets) is increasing with the number of servers s_m in cell m (with the number of pallets N). Finally in (8.2.7.3) the number of machines s_m for each cell and the number of pallets N are limited to positive integers.

Procedure for solution

To solve the above model a two step procedure is proposed. First, with the help of a marginal allocation approach, a heuristic solution is found. Afterwards, in the second step, an implicit enumeration is carried out to find the global optimum.

The marginal allocation approach can be stated as follows:

Greedy heuristic for capacity optimization with budget constraint
Step 0:
set $s_m = 1$ $m = \{1, \dots, M\}$
$\quad N = 1$
$\quad I = \{1, \dots, M+1\}$

Step 1:

calculate $\Delta F_m = \dfrac{T(\underline{s}+\underline{1}_m, N) - T(\underline{s}, N)}{g_m(s_m+1) - g_m(s_m)}$ with $m=\{1,\ldots,M\}$

and $\underline{1}_m$ being a zero vector having the value 1 at position m

$$\Delta F_{M+1} = \dfrac{T(\underline{s}, N+1) - T(\underline{s}, N)}{g_{M+1}(N+1) - g_{M+1}(N)}$$

if $\max_m \Delta F_m \leq 0$ with $m=\{1,\ldots,M+1\}$ go to step 3

set $m^* = \arg(\max_m \Delta F_m)$ with $m=\{1,\ldots,M+1\}$

Step 2:

if $M+1 \neq m^*$

  if $\sum_{\substack{k=1 \\ k \neq m^*}}^{M+1} g_k(s_k, N) + g_m(s_{m^*}+1) \geq BU$ then $I=I\backslash\{m^*\}$

  else add a piece of equipment of type m^* to $\underline{s}$: $\underline{s} = \underline{s}_{m^*} + \underline{1}$

else

  if $\sum_{k=1}^{M} g_k(s_k, N) + g_{M+1}(N+1) \geq BU$ then $I=I\backslash\{m^*\}$

  else add another pallet $N = N + 1$

Step 3:

if $I=\{\}$ or $\max_m \Delta F_m \leq 0$ with $m=\{1,\ldots,M+1\}$ stop: $(\underline{s}, N)$ is the solution

else go to step 1.

alg. 8.4: A greedy heuristic for capacity optimization with budget constraint

After a heuristic solution has been found, an enumeration scheme is applied. To avoid excessive calculations of the throughputs for different configurations, bounds can be provided by:

- the maximal budget BU available compared to the cost of the current configuration $(\underline{s}, N)$,

- and by comparing the throughput T^* of the current best solution (at the beginning of the enumeration the current best solution is the heuristic solution) with the upper bound $UB(\underline{s}, N)$ for the throughput of the current configuration vector $(\underline{s}, N)$.

The upper bound $UB(\underline{s}, N)$ can be estimated through balanced job bound analysis in the following way:[216]

$$T(\underline{s},\underline{w},N) \quad \leq \quad \frac{N \cdot M}{(N+M-1) \cdot \sum\limits_{i=1}^{M} (W_i/s_i)} \quad \equiv \quad UB(\underline{s},N)$$

If the upper bound of the throughput for the configuration under consideration is already less than the best solution found so far, an exact evaluation of the throughput can be omitted.

Example

Based on the part data in chapter 8.1.2.3 (tab.8.1), an example is provided which generates a system for an available budget of $BU = 8,000,000$. Further equipment data is required. Linear cost functions are assumed for the server costs at each cell or the transportation system, i.e. they consist of variable costs IV_m for each machine, vehicle etc. and fixed cell costs or transportation system costs IF_m (tab.8.7):

$$g_m(s_m) = IF_m + IV_m \cdot s_m$$

	L/U St.	mill-1	mill-2	mill-3	vtl-1	vtl-2	transp.
IF_m	10000	50000	100000	100000	100000	50000	500000
IV_m	50000	1000000	800000	1200000	1500000	800000	20000

tab. 8.7: Component data

As further input the routing fractions $\underline{q}$ are given. Here results are used obtained from the adapted FDC algorithm in chapter 8.1.2.3 shown in tab.8.4.

As a result of the algorithm the following configuration is obtained (tab.8.8).

	L/U St.	mill-1	mill-2	mill-3	vtl-1	vtl-2	transp.	pallets
s_m ex.	3	1	1	1	1	1	3	23
s_m heu.	3	1	1	1	1	1	3	23

tab. 8.8: Configuration of the resulting flexible manufacturing system

budget	8,000,000
heur. sol.	7,988,945
exact sol.	7,988,945

tab. 8.9: Available and used budget

216   Zahorjan, J., Sevcik, K.C., Eager, D.L., Galler, B.: Balanced Job Bound Analysis of Queueing Networks, in: Comm. ACM, 25(1982)2, pp.134-141; Eager, D.L., Sevcik, K.C.: Performance Bound Hierarchies for Queueing Networks, in: ACM Trans. Comp. Syst., 1(1983), pp.99-115

With the given vector of flow fractions $\underline{q}$ from the previous chapter 8.1.2.3 the following production rates for each route are obtained:

	part1		part2		part3		
	route1	route2	route1	route2	route1	route2	route3
q_{pr}	0.098	0.188	0.294	0.098	0.311	0.010	–
T_{pr}	1623	3114	4870	1623	5151	166	–

tab. 8.10: Flow fraction on each route

Discussion

Input:
- A limitation for the available budget must be provided.

Model:
- Dynamic system behavior modelling allows the application of this model to flexible machining systems.
- The model is limited to one universal pallet type. A system with several specialized pallet types yielding a different system behavior is not considered.
- It is assumed that there is no restriction concerning the demand for finished parts. This is for example the case when the flexible manufacturing system is the bottleneck of the production process.
- Further limitations are given by the stochastic assumptions of closed queueing network theory.[217]

Procedure:
- The optimal solution is obtained.

8.3 Models for equipment optimization

Models for equipment optimization are capable of selecting the major types of equipment, e.g. CNC-machines, load/unload stations or transportation systems. Furthermore they specify the amount of each type.

8.3.1 Models with an unlimited number of pallets

As shown in chapter 7.2, models with an unlimited number of pallets neglect dynamic system behavior. However, through linearization they allow simpler models to be designed and are quite accurate if a deterministic part flow is given, having only a few part groups

---

217   see chapter 6.1.2.2

with almost no routing-interference, i.e. with flow line characteristics. Furthermore, if the number of pallets in the actual system is high, the throughput behavior of a flexible manufacturing system is almost identical whether modelled statically or dynamically.

8.3.1.1 Equipment selection for flexible assembly systems by Graves and Whitney

The model by Graves and Whitney addresses the design of flexible assembly systems. But, as the authors state, it can be applied just as easily to flexible manufacturing systems.[218]

The objective of the model is to perform the equipment selection and the assignment of assembly-operations in such a way that total costs, consisting of capital costs, operating costs, and costs for load/unload operations, are minimized. This is done under the consideration of annual production requirements for a specific set of tasks.

Model EQUP-Graves/Whitney

Indices:
m   :   index for assembly station
k   :   task index

Decision variables:
δ_m :   binary variable which equals one if station m is included in the system, zero if not.
x_{mk} :   fraction of annual volume of task k assigned to station m.
y_{mk} :   faction of task k that must be loaded onto machine m.

Parameters:
CLU_{mk} :   costs for load/unload operations at machine m for task k
CO_{mk} :   operating costs for task k at machine m
I_m :   capital costs for station m
K_m :   availability level of station m (hours/year)
I_{mk} :   annual load/unload time of task k at machine m
t_{mk} :   annual operating time of task k at machine m

Objective function:

$$\text{Min} \sum_{m=1}^{M} \left(I_m \cdot \delta_m + \sum_{k=1}^{K} (CO_{mk} \cdot x_{mk} + CLU_{mk} \cdot y_{mk}) \right) \qquad (8.3.1.1.1)$$

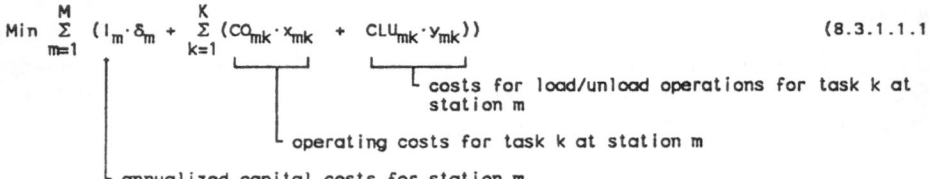

costs for load/unload operations for task k at station m

operating costs for task k at station m

annualized capital costs for station m

218   Graves, S.C., Whitney, D.E.: A Mathematical Programming Procedure for Equipment Selection and System Evaluation in Programming Assembly, in: Proc. 18th IEEE Conf. on Decision and Control, Fort Lauderdale 1979, p.531

Constraints:

$$\sum_{k=1}^{K} (t_{mk} \cdot x_{mk} + l_{mk} \cdot y_{mk}) \leq K_m \cdot \delta_m \qquad \forall\ m \tag{8.3.1.1.2}$$

$$\sum_{m=1}^{M} x_{mk} = 1 \qquad \forall\ k \tag{8.3.1.1.3}$$

$$y_{mk} \geq x_{mk} - x_{mk-1} \qquad \forall\ m,k \tag{8.3.1.1.4}$$

$$\delta_m = 0,1 \qquad \forall\ m$$

$$x_{mk}, \ y_{mk} \geq 0 \qquad \forall\ k \tag{8.3.1.1.5}$$

The system configuration can be generated from a finite number of $m = 1,...,M$ different assembly stations. Every station is characterized by an annualized capital cost, I_m and has a limited annual availability level K_m (hours/year). A finite set of tasks $k = 1,...,K$ is given and each task is a well-defined operation and has an anticipated annual volume associated with it. If such a task k is assigned to a station m, t_{mk} units of annual operation time is required and specific annual operation costs CO_{mk} are incurred. For each assembly task a load/unload operation can be assigned. However, such a load/unload operation is not performed when the prior task k-1 is being performed on the same machine m. In this case the time l_{mk} and costs CLU_{mk} caused by the load/unload operation for task j have to be zero.

The objective function of the model minimizes annual costs for the system. These costs comprise annual operating costs, load/unload operating costs and annual costs for the selected equipment. The first set of constraints (8.3.1.1.2) ensures that the availability of each station is not exceeded. The next constraint set (8.3.1.1.3) ensures that the annual volume of each task is completely assigned. The constraints in (8.3.1.1.4) and (8.3.1.1.5) serve as a definition for the variable y_{mk}. If x_{mk} is smaller than x_{mk-1}, this means that the production volume at station m for operation k is less than for the prior operation k-1, then y_{mk} becomes zero. If the contrary is the case and x_{mk} is larger than x_{mk-1}, then y_{mk} equals the difference $x_{mk}-x_{mk-1}$.

Procedure for solution

The model is solved by a branch-and-bound procedure. In this context it means an implicit enumeration of all possible combinations of the binary variables. Bounds are evaluated by exploring a Lagrangian relaxation of the model. This is done by dynamic programming. The Lagrangian multipliers are estimated via subgradient optimization.[219]

219 Graves, S.C., Whitney, D.E.: A Mathematical Programming Procedure for Equipment Selection and System Evaluation in Programming Assembly, in: Proc. 18th IEEE Conf. on Decision and Control, Fort Lauderdale 1979, S.532-533; Note that in the presented Lagrangian Function (6) the right hand side of the relaxed constraint is missing.

Discussion

Model:
- Because of static system behavior modelling, the model is restricted to flexible transfer lines and multi-lines or to flexible manufacturing systems with a large number of pallets, where costs dependent on the number of pallets are negligible.
- The model only considers constant demand. This implies, that changing levels in production requirements during the existence of such a flexible assembly system are not anticipated.
- The cell structure of a flexible manufacturing or assembly system is not considered in the cost model. Therefore cell costs are not included.[220]
- Possible cost reductions through the performance of a certain combination of operations are neglected.
- Production criteria other than satisfying demand - e.g. throughput time - cannot be included.
- Pallet, fixture and inprocess inventory costs are neglected.
- Possible machine breakdowns are not explicitly included in the model. However, machine availability K_m may be adjusted downward to reflect the expected annual downtime for the machine.

Procedure:
- The optimal solution is obtained.
- Standard software can only be applied for small problem sizes.

8.3.1.2 *Equipment selection for flexible assembly systems by Graves and Lamar*

The model by Graves and Lamar is an extension of the earlier work by Graves and Whitney in the previous chapter. The following changes were made:
- Now the formulation of the model consists of only binary variables. The earlier model by Graves and Whitney embodies a mixed integer formulation. Consequently that model made it possible to split the annual production volume for one type of operation and to assign its fractions to several stations. This model only allows the assignment of one type of operation to one distinct station. Graves and Lamar justify this restriction with the growing complexity, if not impossibility of generating the physical layout.[221]
- By the introduction of a supplemental index for the operating time and -cost, the inclusion of tool change times and -costs is made.

---

220   see chapter 7.5 for an alternative approach
221   Graves, S.C., Lamar, B.W.: A Mathematical Programming Procedure for Manufacturing System Design and Evaluation, in: Proc. IEEE Int. Conf. on Cir. and Comput., 1980, p.1147

Model EQUP-Graves/Lamar

Indices:

m	:	index for assembly station
j	:	index for task j
k	:	index for following task at a station

Decision variables:

x_{mjk}	:	binary variable equals one if task k follows task j at station m.
y_m	:	binary variable denoting the inclusion of station m in the assembly system configuration.

Parameters:

CO_{mjk}	:	operating costs of task j at station m when task k with $k>j$ is also performed on station m
I_m	:	capital costs for station m
K_m	:	availability level of station m (hours/year)
t_{mjk}	:	operating time of task j at station m when task k with $k>j$ is also performed on station m

Objective function:

$$\min \sum_{m=1}^{M} \left(I_m \cdot y_m + \sum_{j=0}^{J} \sum_{k=j+1}^{J+1} CO_{mjk} \cdot x_{mjk} \right) \qquad (8.3.1.2.1)$$

operating costs of task j at station m when task k with k>j is also performed on station m

annualized capital costs for station m

Constraints:

$$\sum_{m=1}^{M} \sum_{k=j+1}^{J+1} x_{mjk} = 1 \qquad \forall\ j \in J \qquad (8.3.1.2.2)$$

$$\sum_{j=0}^{J} \sum_{k=j+1}^{J+1} t_{mjk} \cdot x_{mjk} - K_m \leq 0 \qquad \forall\ m \in M \qquad (8.3.1.2.3)$$

$$\sum_{k=j+1}^{J+1} x_{mjk} - \sum_{l=0}^{j-1} x_{mlj} = 0 \qquad \forall\ j \in J,\ m \in M \qquad (8.3.1.2.4)$$

$$\sum_{k=1}^{J} x_{mOk} - y_m = 0 \qquad \forall\ m \in M \qquad (8.3.1.2.5)$$

$$\sum_{l=1}^{J} x_{mlJ+1} - y_m = 0 \qquad \forall\ m \in M \qquad (8.3.1.2.6)$$

$$x_{mjk} = 0,1 \qquad y_{mjk} = 0,1 \qquad \forall\ m,j,k \qquad (8.3.1.2.7)$$

The set of disposable stations from which to choose is given by M. For one station m there are annual costs of I_m and an availability level of K_m. The availability level reflects, as in the model of the previous chapter, the expected operating time per year for the station m (hours/year). The set of distinct assembly tasks which the system must perform, is given by J. The operating time is defined as the total annual operating time t_{mjk} required to perform task j at station m if the following task k $(k > j)$ is also performed at station m. In detail the operating time t_{ijk} can consist of:

- the actual assembly time associated with task j,
- tool change time if task k requires a different tool than that used for the previous task j,
- and a possible load/unload time resulting from the fact, that task k (with $k > j$) is the next operation done at station m.

Note that there is no load/unload time, if station m can perform task k immediately after performing task j on a particular assembly or part.[222]

If there is no operation prior to k, i.e. k is the first operation at station m, then t_{m0k} is the total additional operating time associated with task k. On the other hand, if j is the last operation at station m, then the time t_{mjJ+1} comprises total operating time associated with task j. Analog to the definition of the operating times is the definition of the operating costs CO_{mjk}. They include the following costs:

- variable operating costs
- variable tooling costs
- variable handling costs.

The objective function minimizes annual capital costs for equipment and annual operating costs as a function of the decision variables mentioned above. The first set of constraints (8.3.1.2.2) ensures that each task is assigned to a work station, whereas constraint set (8.3.1.2.3) limits the amount of work assigned to a particular station m to its availability level K_m. Constraints in (8.3.1.2.4) enforce the sequential precedence relationships of all tasks at each station. This signifies that at station m the subsequent operation k of operation j can only be of type $k > j$. Constraint set (8.3.1.2.5) in conjunction with (8.3.1.2.6) and (8.3.1.2.4) relate the station selection variables y_m to the task assignment variables x_{mjk}. If, for example, station m is not included in the configuration ($y_m = 0$), all x_{mjk} must be enforced to be zero.

Procedure for solution

The model formulation consists of a large binary integer program. Thus Graves and Lamar only consider an approximate solution procedure. This is done by a reformulation of the problem and a relaxation of the integer constraints for the new decision variables. By solving the relaxed version, lower bounds are obtained. The procedures used are generalized linear programming, or, alternatively, methods similar to those used by Graves and Whitney in the preceding model. Upper bounds are obtained through feasible integer solutions derived as a by-product of the previous step in which lower bounds were established. The best upper bound provides the approximate solution. The generated lower bound yields a conservative a posteriori measure on the near-optimality of the solution.[223]

---

222 Graves, S.C., Lamar, B.W.: A Mathematical Programming Procedure for Manufacturing System Design and Evaluation, in: Proc. IEEE Int. Conf. on Cir. and Comput., 1980, p.1147

223 for a more detailed description of the algorithm see: Graves, S.C., Lamar, B.W.: A Mathematical Programming Procedure for Manufacturing System Design and Evaluation, in: Proc. IEEE Int. Conf. on Cir. and Comput., 1980, p.1148-1149; Graves, S.C., Lamar, B.W.: An Integer Programming Procedure for Assembly System Design Problems, in: OR, 31(1983)3, p.531-537

Discussion

Model:

- Because of static system behavior modelling, the model is restricted to flexible transfer lines and multi-lines or on flexible manufacturing systems with a large number of pallets, where costs dependent on the number of pallets are negligible.

- The model only considers constant demand. This implies, that changing levels in production requirements during the existence of such a flexible assembly system are not anticipated.

- Tool costs or tool change times are considered.

- The cell structure of a flexible manufacturing or assembly system is not considered in the cost model. Therefore cell costs are not implemented.[224]

- Possible cost reductions through the performance of a certain combination of operations are neglected.

- Production criteria other than satisfying demand - e.g. throughput time - cannot be included.

- Possible machine breakdowns are not explicitly included in the model. However, machine availability K_m may be adjusted downward to reflect the expected annual downtime for the machine.

- Splitting of the annual workload of one operation over several machines is not possible.

- Pallet, fixture and inprocess inventory costs are neglected.

Procedure:

- Only an approximately optimal solution is obtained. However a conservative a posteriori measure on the near-optimality of the solution is given.

8.3.1.3 A new model for optimal equipment selection

Below a new model for equipment optimization is presented which is intended to minimize annual equipment and operating costs. The main difference between it and the preceding models for equipment optimization is that it makes more precise allowance for the cell structure of flexible manufacturing systems. It includes the following equipment:

- cells with one or more identical CNC-machines, set-up tables or load/unload stations,

- transportation systems with one or more vehicles.

Here the optimization model refers to the cost model introduced in chapter 7.5. Two possible ways of considering cell costs are given. Firstly in a more general setting, the cell costs or transportation system costs can be decribed by an increasing concave cost function in the number of machines, load/unload stations, vehicles etc.. Secondly a more restricted modelization is given, if for each cell linear cost functions with variable costs depending on the number of identical servers (CNC-machines, load/unload stations etc.) and fixed costs

---

224   see chapter 7.5 for an alternative approach

are assigned. These fixed costs allow, for instance, costs for a local material or tool handling systems to be included into the optimization process. Similarly, variable costs are assigned to the transportation system depending on the number of carriers or vehicles implemented. Here fixed costs allow costs for the computer hardware etc. to be considered.

Depending on the selection of equipment, a routing optimization is made so that total costs, i.e. annual equipment and operating costs, are minimized. The routing optimization is based on the selection of alternative routes for a part group, i.e. different alternative sets of workloads at cells used for processing.

Model EQUP-NEW-(1)

Indices:

b :   index for number of servers (machines, set-up tables, carriers etc.) at cell
g :   part group index
m :   index for cells or transportation system
r :   index for routes

Decision variables:

z_{bm} :   binary integer variable which is one if cell (transportation system) m with the amount of b machines (vehicles) is included in the configuration of the flexible manufacturing system.
x_{gr} :   production rate of part group g produced on route r

Parameters:

EF_m :   efficiency factor for equipment of type m
CO_{gr} :   operating costs for part group g on route r
$frac_g$:   product rate, i.e. the fraction part group g has of total production
I_{bm} :   equipment costs for cell or transportation system having b machines or respectively b vehicles
R_{min} :   required production rate for the system
W_{gmr} :   workload of part group g at cell m on route r

Objective function:

$$\min \sum_{m=1}^{M} \sum_{b=1}^{B_m} I_{bm} \cdot z_{bm} \; + \; \sum_{g=1}^{G} \sum_{r=1}^{R_g} CO_{gr} \cdot x_{gr} \qquad\qquad (8.3.1.3.1)$$

              ⎣_____⎦   ⎣_____⎦
         annual equipment costs   annual operating costs

Constraints:

$$\sum_{g=1}^{G} \sum_{r=1}^{R_g} x_{gr} \cdot W_{gmr} \; \leq \; \sum_{b=1}^{B_m} b \cdot z_{bm} \cdot EF_m \qquad \forall \; m \qquad (8.3.1.3.2)$$

                          ⎣__ efficiency factor
                      number of machines/transport vehicles of type m

        ⎣__ workload of part group g at cell m on route r

     ⎣__ production rate of part group g on route r

$$\sum_{r=1}^{R_g} x_{gr} = frac_g \cdot R_{min} \qquad \forall\, g \qquad\qquad (8.3.1.3.3)$$

production rate required

product rate of part group g

production rate for part group g on route r

$$x_{gr} \geq 0 \quad \forall\, g,r \qquad z_{bm} = \{0,1\} \quad \forall\, m,b \qquad\qquad (8.3.1.3.4)$$

In the objective function annual equipment costs and annual operating costs are minimized. Cell costs or costs for the transportation system depend on the number of machines, load/unload stations, vehicles etc. given by the index b. These costs are increasing concave, when b is increasing. The annual operating costs are obtained by the sum over all operating costs CO_{gr} for each part group g produced on the alternative routes r multiplied with the annual production rate x_{gr}. Compared with the new routing models presented in chapter 8.1.2.3 and 8.1.2.4, where the routing fractions q were the decision variables, here the amount of parts of each part group produced in one period on each alternative route is used. Both variables are related through the following expression:

$$x_{gr} \Big/ \sum_{g=1}^{G} \sum_{r=1}^{R_g} x_{gr} = q_{gr}$$

Constraint set (8.3.1.3.2) is a capacity constraint equivalent to the one used in the models for static system behavior modelling in the previous chapters.[225] Note that an efficiency factor EF_m can be introduced for each type of cell/transportation system to reduce its utilization and thus to approximate dynamic system behavior and machine breakdowns. To guarantee the realization of a minimum production rate for every part group, constraint set (8.3.1.3.3) is introduced. Hence the product of the minimum production rate R_{min} with the product rate $frac_g$ must be equal to the production over all routes for one part group g.

Procedure for solution

The model presented above consists of a mixed integer program. In the following three solution procedures are presented.

The first one will be given by a branch-and-bound procedure, where the integer constraints are relaxed for the evaluation of lower bounds. Furthermore a reformulation of the model is suggested, to restrict branching.

Alternatively a branch-and-bound procedure is presented for the original model formulation which includes a relaxation of the capacity constraint using a Lagrange function. It allows the evaluation of lower bounds more quickly and thus accelerates the bounding process.

Finally a simple greedy heuristic to quickly obtain a good and feasible solution is provided. The latter can also be used as a starting vector or an upper bound for the other two procedures.

---

225   see chapter for example chapter 8.1.1.1 or 8.1.1.2

Branch-and-bound with relaxed integer constraints

The above model can be solved by a standard code for mixed integer models using a branch-and-bound type of algorithm in conjunction with the simplex method to obtain lower bounds. However, the use of such a branch-and-bound code for mixed integer problems can lead to certain disadvantages:[226]

- Extensive branching is required just to reach a feasible solution.

- Fathoming many near-optimal solutions due to minor shifts in capacity can be very tedious in the branch-and-bound procedure.

To avoid extensive branching a reformulation of EQUP-NEW-(1) is suggested which, assuming linear cost functions for the cells, reduces the number of integer variables considerably. Furthermore a feasible solution can be provided at the beginning of the algorithm by a greedy heuristic, which will be explained in more detail later.

Indices:

b	:	index for number of servers (machines, set-up tables, carriers etc.) at cell
g	:	part group index
m	:	index for cells or transportation system
r	:	index for routes

Decision variables:

s_m	:	integer variable for the number of machines, load/unload stations, vehicles etc. at a cell or the transportation system of type m
y_m	:	binary variable which is one if cell (machine type) or the transportation system of type m is included into the configuration of the flexible manufacturing system
x_{gr}	:	production rate of part p produced on route r

Parameters:

EF_m	:	efficiency factor for equipment of type m
CO_{gr}	:	operating costs for part group g on route r
$frac_g$	:	product rate, i.e. the fraction part group g has of total production
IF_m	:	fixed cell costs or fixed costs for the transportation system
IV_m	:	variable cell costs or variable costs for the transportation system for each additional machine, load/unload station, vehicle etc.
LN	:	large number
R_{min}	:	required production rate for the system
W_{gmr}	:	workload of part group g at cell m on route r

Objective function:

$$\min \sum_{m=1}^{M} (IF_m \cdot y_m + IV_m \cdot s_m) \; + \; \sum_{g=1}^{G} \sum_{r=1}^{R_g} CO_{gr} \cdot x_{gr} \qquad (8.3.1.3.4)$$

$$\underbrace{\phantom{\sum_{m=1}^{M} (IF_m \cdot y_m + IV_m \cdot s_m)}}_{\text{annual equipment costs}} \qquad \underbrace{\phantom{\sum_{g=1}^{G} \sum_{r=1}^{R_g} CO_{gr} \cdot x_{gr}}}_{\text{annual operating costs}}$$

226   Erlenkotter, D.: A comparative study of approaches to dynamic location problems, in: EJOR, 6(1981), p.135

Constraints:

$$\sum_{g=1}^{C} \sum_{r=1}^{R_g} x_{gr} \cdot W_{mgr} \leq s_m \cdot EF_m \qquad \forall\, m \tag{8.3.1.3.5}$$

    ⌐ efficiency factor for machines/transport vehicles of type m
   ⌐ number of machines/transport vehicles of type m
  ⌐ workload of part group g at machine type m on route r
⌐ production rate of part group g on route r

$$\sum_{r=1}^{R_g} x_{gr} = frac_g \cdot R_{min} \qquad \forall\, g \tag{8.3.1.3.6}$$

   ⌐ required production rate
  ⌐ product rate of part group g
⌐ production rate for part group g on route r

$$s_m \leq y_m \cdot LN \qquad \forall\, m \text{ and } LN \text{ large number} \tag{8.3.1.3.7}$$

$$x_{gr} \geq 0 \;\; \forall\, g,r \qquad s_m \in Z^+_0 \;\; \forall\, m \qquad y_m = \{0,1\} \;\; \forall\, m \tag{8.3.1.3.8}$$

   In the objective function cell costs or costs for the transportation system now consist of a variable component IV_m which depends on the number of machines, load/unload stations, set-up tables or vehicles selected, and a fixed component IF_m. A further constraint set (8.3.1.3.7) is introduced to ensure that if $s_m > 0$ then $y_m = 1$. Cost minimization causes the annual cell or transportation system costs for type m to be zero if there is no machine, set up-table or vehicle of this type in the system. Conversely, if there is equipment of type m integrated in the system, the fixed costs will be added to the variable machine costs in the objective function.

Branch-and-Bound with Lagrangian relaxation

   Another approach to solve model EQUP-NEW-(1) exactly is given by a Lagrangian relaxation embedded in a branch-and-bound process. In this way lower bounds - and therefore convergence results - can be obtained more quickly. The Lagrangian relaxation is carried out by introducing the capacity constraints into the objective function.

$$LF(\underline{\alpha}) = \min_{\underline{z},\underline{x}} \sum_{m=1}^{M} \sum_{b=1}^{B_m} I_{bm} \cdot z_{bm} + \sum_{g=1}^{C} \sum_{r=1}^{R_g} CO_{gr} \cdot x_{gr}$$
$$+ \sum_{m=1}^{M} \alpha_m \cdot \left(\sum_{g=1}^{C} \sum_{r=1}^{R_g} x_{gr} \cdot W_{gmr} - \sum_{b=1}^{B_m} b \cdot z_{bm} \right) \tag{8.3.1.3.9}$$

Constraints:

$$\sum_{r=1}^{R_g} x_{gr} = frac_g \cdot R_{min} \qquad \forall\, g \tag{8.3.1.3.10}$$

$$x_{gr} \geq 0 \;\; \forall\, g,r \qquad z_{bm} = \{0,1\} \;\; \forall\, m,b \tag{8.3.1.3.11}$$

For a given solution vector $(\underline{z},\underline{x})$ the Lagrangian function has to be maximized to get a tight lower bound on the objective function $F(\underline{z},\underline{x})$:

$$F(\underline{z},\underline{x}) \geq \max_{\underline{\alpha}} LF(\underline{\alpha}) \quad \text{with } \alpha_m \geq 0 \; \forall \; m \qquad (8.3.1.3.12)$$

To maximize the Lagrangian Function over $\underline{\alpha}$, the subgradient approach can be used, based on the work of Held and Karp[227] and Held, Wolfe and Crowder[228]. The Lagrange multipliers $\underline{\alpha}$ are obtained through the following recursive formula:

$$\alpha_m^{k+1} = \alpha_m^k + a^k \cdot (\sum_{g=1}^{G} \sum_{r=1}^{R_g} x_{gr}^k \cdot W_{gmr} - \sum_{b=1}^{B_m} b \cdot z_{bm}^k) \qquad (8.3.1.3.13)$$

$$\text{with} \quad a^k = \frac{\lambda^k \cdot (LF^* - LF(\underline{\alpha}^k))}{\left| (\sum_{g=1}^{G} \sum_{r=1}^{R_g} x_{gr}^k \cdot W_{gmr} - \sum_{b=1}^{B_m} b \cdot z_{bm}^k) \right|^2}$$

Initial values for α_m^0 can be provided by I_{1m}. The factor LF^* is an upper bound on $LF(\underline{\alpha}^k)$ and the factor λ_k is a scalar between $0 < \lambda_k \leq 2$. The sequence λ_k is determined by setting $\lambda_0 = 2$ and halving λ_k whenever $LF(\underline{\alpha}^k)$ has failed to increase after ΣB_m iterations.[229]

However, to obtain the value of the Lagrangian function, first a solution $(\underline{z},\underline{x})$, which minimizes the above model, has to be generated. This can be done separately for $\underline{z}$ and $\underline{x}$. First, to obtain values for $\underline{z}$ at positions not yet fixed by the branch-and-bound procedure, their influence on the objective function is examined. For given Lagrange Multiplier α_m minimization of the Lagrangian function can only be achieved by the following rules:

$$\begin{array}{lll} z_{bm} = 1 & \text{if } I_{bm} \cdot z_{bm} - \alpha_m \cdot b \cdot z_{bm} < 0 & (8.3.1.3.14) \\ z_{bm} = 0 & \text{if } I_{bm} \cdot z_{bm} - \alpha_m \cdot b \cdot z_{bm} > 0 & (8.3.1.3.15) \\ z_{bm} = 0,1 & \text{if } I_{bm} \cdot z_{bm} - \alpha_m \cdot b \cdot z_{bm} = 0 & (8.3.1.3.16) \end{array}$$

Because the cost function is increasing concave, only that z_{bm} is set to one which fulfills inequality (8.3.1.3.14) with the highest b. Thus inequality (8.3.1.3.14) is changed to:

$$\text{for max } (\{b \mid I_{bm} \cdot z_{bm} - \alpha_m \cdot b \cdot z_{bm} < 0\}) \qquad (8.3.1.3.14')$$
$$\text{set } z_{bm} = 1 \text{ else } z_{bm} = 0$$

The values of $\underline{x}$ can be obtained by solving the following flow problem:

$$\min_{\underline{x}} \sum_{g=1}^{G} \sum_{r=1}^{R_g} [(\sum_{m=1}^{M} \alpha_m \cdot W_{gmr}) + CO_{gr}] \cdot x_{gr} \qquad (8.3.1.3.17)$$

Constraints:

$$\sum_{r=1}^{R_g} x_{gr} = frac_g \cdot R_{min} \quad \forall \; g \qquad (8.3.1.3.18)$$

227   Held, M., Karp, R.M.: The traveling salesman problem and minimum spanning trees, in: Mathematical Programming, 1(1971), pp.6-25
228   Held, M., Wolfe, P., Crowder, H.D.: Validation of subgradient optimization, in: Mathematical Programming, 6(1974), pp.62-88
229   Fisher, M.L.: The Lagrangian Relaxation Method for solving Integer Programming Problems, in: Management Science, 27(1981)1, p.8

$$x_{gr} \geq 0 \quad \forall \, g,r \qquad\qquad (8.3.1.3.19)$$

This can easily be solved by assigning the flow $frac_g \cdot R_{min}$ on that route of each part group g which has the least value for the term in square brackets of the objective function.

For a detailed description on the implementation of the Lagrangian relaxation into the branch-and-bound procedure see Fisher et al.[230], Shapiro[231] and Geoffrion[232].

Greedy heuristic

The procedure assigns in steps the required production rate of each part group. According to the general description of greedy algorithms the heuristic chooses the best of all available candidates at each step without considering the consequences on future steps (see alg.7.3).[233] Production flows on the routes r of all part groups g are used as the set of useable candidates. Furthermore there is a selection function, which selects one of the candidate routes at each step according to the following cost factor:

$$\min_{\substack{g,r \\ SR_g > 0}} \left[\underbrace{\underbrace{\sum_{m \in R_g} (CO_{mgr} + \sigma_m + \beta_m)}_{\text{actual equipment costs per part}}}_{\text{operating costs per part}} \right]$$

$$\text{with} \quad \sigma_m = \begin{cases} IF_m \cdot W_{mgr}/K & \text{if } SMW_m = 0 \\ 0 & \text{if } SMW_m > 0 \end{cases}$$

$$\text{and} \quad \beta_m = \begin{cases} IV_m \cdot W_{mgr}/K & \text{if } SMW_m = 0 \text{ or integer} \\ 0 & \text{if } SMW_m \text{ noninteger} \end{cases}$$

The costs for each route are calculated in such a way that operating costs CO_{mgr} and equipment costs are included. Note that linear cost functions are used for equipment costs. However, any arbitrarily increasing cost functions could be used instead. Fixed IF_m and variable equipment costs IV_m are considered for the cost factor if the cell (transportation system) was not yet considered, i.e. the required machine capacity SMW_m is zero. When the required machine capacity SMW_m of the cell (transportation system) is a positive integer, i.e. it is fully utilized and has no more free capacity, only variable equipment costs IV_m are included. If neither of both cases is true, free capacity of cell type m or the transportation system of type m is available and thus its equipment costs are zero. By dividing equipment cost through the production time K per period and multiplying it by the processing time, the given factor selects that route which has the minimum costs

230   Fisher, M.L., Northup, W.D., Shapiro, J.F.: Using Duality to solve Discrete Optimization Problems: Theory and Computational Experience, in: Mathematical Programming Study, 3(1975), pp.76-79

231   Shapiro, J.F.: A Survey of Lagrangian Techniques for Discrete Optimization, in: Annals of Discrete Mathematics, 5(1979), pp.132-135

232   Geoffrion, A.M.: Lagrangian Relaxation for Integer Prgramming, in: Mathematical Programming Study, 2(1974), pp.82-114

233   For a description of greedy algorithms see chapter 6.1.1.3

(equipment and operation costs) for the production of one part.

Next a feasibility function checks whether the part group of the selected route still shows production requirements, i.e. $SR_g > 0$.

If the production rate for the part group of the selected route is already achieved, all routes of that part group are removed from the set of useable candidates and a new route is selected. Otherwise production flow is assigned to the route with the lowest cost factor until the first machine is fully utilized:

$$\min_{\substack{x \\ g^*r^*}} \{ x_{g^*r^*} \cdot W_{g^*mr^*} / EF_m = [SMW_m] - SMW_m \} \quad \forall m \quad \text{with } x_{g^*r^*} \leq SR_g \qquad 234$$

remainder of free capacity at cell m
of the current configuration

possible throughput x_{gr} until capacity limit at cell m is reached

This is because a new machine is then required and consequently there is a change in the cost factor. In the above formulation the part flow multiplied by the workload W_{mpr} and adjusted by the efficiency factor EF_m is less than or equal to the available remainder of capacity, expressed by the right-hand side of the equality.

Afterwards the route r^* with its production flow x is assigned to the set of selected candidates.

The algorithm proceeds until the production requirements for all parts are fulfilled, i.e. a final solution has been found. The heuristic solution is then given by the part flow provided by the set of selected candidates sx_{gr} and the machine requirements necessary to produce this part flow.

Greedy heuristic for EQUP-NEW-(1)

Step 0:

initialize $SMW_m = 0 \quad \forall m$, $SR_g = frac_g \cdot R_{min}$

and $\qquad sx_{gr} = 0 \quad \forall g, r$

---

234   The variable in square brackets [ ] is rounded off to the next higher integer value.

Step 1:

$$\text{find} \quad \min_{\substack{g,r \\ SR_g > 0}} \left[\sum_{m \in R_g} (CO_{mgr} + \sigma_m + \beta_m) \right]$$

with $\sigma_m = \begin{cases} IF_m \cdot W_{mgr} / K & \text{if } SMW_m = 0 \\ 0 & \text{if } SMW_m > 0 \end{cases}$

and $\beta_m = \begin{cases} IV_m \cdot W_{mgr} / K & \text{if } SMW_m = 0 \text{ or integer} \\ 0 & \text{if } SMW_m \text{ noninteger} \end{cases}$

Step 2:

find $\min_{x_{g^*r^*}} \{ x_{g^*r^*} \cdot W_{g^*mr^*} / EF_m = [SMW_m] - SMW_m \} \ \forall \ m$ with $x_{g^*r^*} \leq SR_g$ [235]

Step 3:

set $SMW_m = SMW_m + x_{g^*r^*} \cdot W_{gmr} / EF_m$, $sx_{gr} = sx_{gr} + x_{g^*r^*}$

and $SR_g = SR_g - x_{g^*r^*}$

Step 4:

if $SR_g = 0 \ \forall \ g$

    then stop: sx_{gr} equals the production flow of each route,

      the value of SMW_m rounded to the next higher integer value yields the number

      of machines/set-up tables/vehicles of type m: $s_m = [SMW_m]$;

else go to step 1.

---

alg. 8.5: Greedy heuristic for EQUP-NEW-(1)

Example

Use of the above model is demonstrated in the following example. It likewise consists of a part family of 3 different part types {part1, part2, part3}. For each part type different alternative routes exist, i.e. part1 and part2 have two, and part3 has three alternative routes. Each single part type forms hereby a part group. For the operating times refer to tab.8.1 in chapter 8.1.2.3 and for the operating costs to tab.8.5 in chapter 8.1.2.4. The

---

[235] The variable in square brackets [ ] is rounded off to the next higher integer value.

equipment costs and efficiency factors EF_m are given in tab.8.11. Linear cost functions are assumed for the server costs at each cell or the transportation system, i.e. they consist of variable costs IV_m for each machine, vehicle etc. and fixed cell costs or transportation system costs IF_m (tab.8.11):

$$g_m(s_m) = IF_m + IV_m \cdot s_m$$

	L/U St.	mill-1	mill-2	mill-3	vtl-1	vtl-2	transp.
IF_m	5000	25000	50000	50000	50000	25000	250000
IV_m	25000	500000	400000	600000	750000	400000	10000
EF_m	1.00	1.00	1.00	1.00	1.00	1.00	1.00

tab. 8.11: Component data

Finally input data concerning the production requirements R_g for each part type for an average period must be given (tab.8.12).

	part1	part2	part3
R_g	11000	15000	12000

tab. 8.12: Production requirements

As the result of the greedy heuristic the following configuration (tab.8.13) and part flows (tab.8.14) can be obtained. They are compared with the exact solution obtained through branch-and-bound. The value of the objective function of the heuristic solution is 8,640,020. The exact solution with 6,956,110 is about 19% lower than the heuristic solution.

	L/U St.	mill-1	mill-2	mill-3	vtl-1	vtl-2	transp.
s_m (heur.)	2	3	4	3	2	0	2
s_m (exact)	3	2	2	1	2	2	3

tab. 8.13: Resulting configuration of the flexible manufacturing system

	part1		part2		part3		
	route1	route2	route1	route2	route1	route2	route3
T (heu.)	11000	–	–	15000	–	–	12000
T (ex.)	9850	1150	11700	3310	12000	45	–

tab. 8.14: Production per period on each route

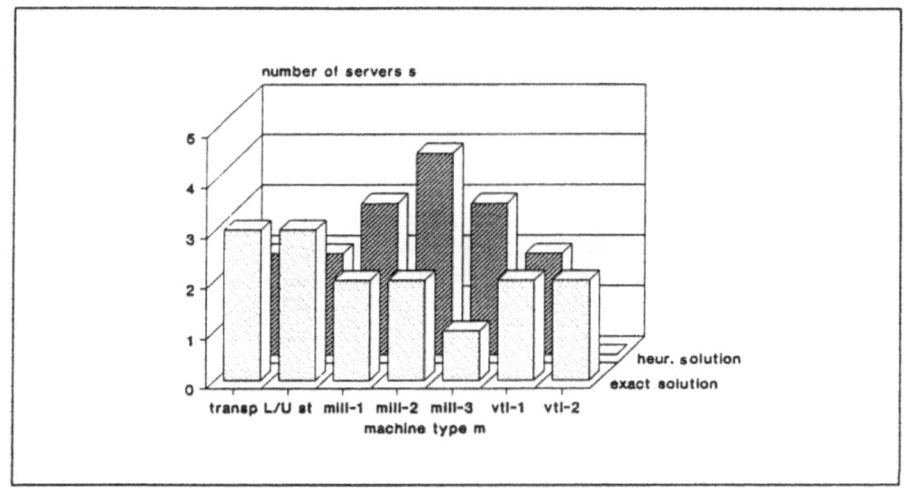

fig. 8.5: The configuration of the flexible manufacturing system

Discussion

Input:
- Different alternative routes must be provided.

Model:
- Because of static system behavior modelling, the model is restricted to flexible transfer lines and multi-lines or to flexible manufacturing systems with a large number of pallets, where costs dependent on the number of pallets are negligible.
- The model only considers constant demand. This implies, that changing levels in production requirements are not anticipated during the existence of a system.
- The cell structure of a flexible manufacturing system is considered in the cost model.
- Possible cost reductions through the performance of a certain combination of operations are neglected.
- Production criteria other than satisfying demand - e.g. throughput time - cannot be included.
- Possible machine breakdowns are not explicitly included in the model. However, an efficiency factor is used which allows the expected annual downtime for a cell to be reflected approximately.
- Pallet, fixture and inprocess inventory costs are neglected.

Procedure:
- The optimal solution can be obtained with any commercial software package for mixed

integer programming.

- For the model formulation with linear cost functions of cell and transportation system costs practical problem sizes can be solved in acceptable time with standard branch-and-bound algorithms. Similar results are obtained for the model formulation with general increasing concave costs functions using Lagrangian relaxation embedded in a branch-and-bound procedure.[236]

- The greedy heuristic allows the derivation of an upper bound useable at the beginning of branch-and-bound algorithms. A sample of tests runs with different numbers of alternative machine types, routes and part types showed an average deviation from optimality of 19%. A documentation of these test runs can be found in Appendix D.

8.3.1.4 Extension: considering flexibility aspects

Most often product life cycles are shorter than the lifetime of a flexible manufacturing system. This causes qualitative (part groups) and quantitative (production volume) changes of the part family over the system's lifetime, which do not allow the configuration of a flexible manufacturing system to be based on an average period. Moreover, changing levels in production requirements make it necessary to determine how much volume and how much expansion flexibility the system should possess.[237] For instance, the question arises whether the system should have some overcapacity at the beginning, or if it is better to expand it later, when the capacity is required. Should these changes in production volume and part types occur, the system designer has to incorporate a dynamic feature into his decision model to evaluate the optimal supply in volume, expansion and production flexibility. It is suggested that the optimal supply of these flexibilities can be derived under the assumption of a production demand with certainty, by simply introducing a time index into the EQUP-NEW-(1) model and dividing the system's lifetime into a set of time periods each having constant demand. Furthermore, discounted cash outflows, i.e. discounted payments, instead of costs will be minimized. Optimizing the configuration of a system then allows the system to be adapted to the requirements of each period in such a way that the present value of the cash outflows are optimized over the system's lifetime.

Model EQUP-NEW-(2)

Indices:
b	:	index for number of servers (machines, set-up tables, carriers etc.) at cell
g	:	part group index
m	:	index for cells or transportation system
r	:	index for routes
t	:	time index

Decision variables:
z_{bmt}	:	binary integer variable, which is one if cell (transportation system) m with the number of b machines (vehicles) is included in the configuration of the flexible manufacturing system in period t.

---

236   see results in Appendix D
237   For a more in-depth discussion on this subject, see chapter 7.3

x_{rgt} : production rate of part group g produced on route r in period t

Parameters:

EF_m : efficiency factor for equipment of type m
CF_{grt} : discounted payments for operating a single part of machine group g on route r in period t
CO_{mpr} : payments for operating of part type p on route r at machine m
EF_m : efficiency factor for machines/transport vehicles of type m
F_{bmt} : discounted payments for equipment (cell or transportation system) of type m in period t consisting of b servers (machines, load/unload stations, vehicles etc.)
FFE_m : payments for equipment expansion of type m cell or transportation system
FF_m : payments for equipment of a type m cell or transportation system in the first period
$frac_{gt}$: product rate, i.e. the fraction part group g has of total production in period t
FVE_m : payments for equipment expansion of type m server (machines, load/unload stations, vehicles etc.)
FV_m : payments for equipment of a type m server (machines, load/unload stations, vehicles etc.) in the first period
i : rate of return
R_{tmin} : required production rate for the system in period t
W_{mgr} : workload of part group g on route r at cell (transportation system) m

Objective function:

$$\min \sum_{t=0}^{T} \sum_{m=1}^{M} \sum_{b=1}^{B_m} F_{bmt} \cdot z_{bmt} \; + \; \sum_{t=0}^{T} \sum_{g=1}^{G} \sum_{r=1}^{R_g} CF_{grt} \cdot x_{grt} \qquad (8.3.1.4.1)$$

$\underbrace{\hspace{3cm}}$ discounted payments for equipment

$\underbrace{\hspace{3cm}}$ discounted payments for operating

Constraints:

$$\sum_{g=1}^{G} \sum_{r=1}^{R_g} x_{grt} \cdot W_{mgr} \; \leq \; EF_m \cdot \sum_{u=0}^{t} \sum_{b=1}^{B_m} b \cdot z_{bmu} \qquad \forall \; m,t \qquad (8.3.1.4.2)$$

number of machines/transport vehicles of type m in period t

efficiency factor for machines/transport vehicles of type m

workload of part group g at machine type m on route r

production rate of part group g on route r in period t

$$\sum_{r=1}^{R_g} x_{grt} \; = \; frac_{gt} \cdot R_{tmin} \qquad \forall \; g,t \qquad (8.3.1.4.3)$$

production rate required in period t

product rate of part group g in period t

production rate of part group g on route r in period t

$$x_{grt} \geq 0 \quad \forall \; t,g,r \qquad z_{bmt} = \{0,1\} \quad \forall \; t,m,b \qquad (8.3.1.4.4)$$

Procedure for solution

The model can be solved with any standard algorithm for mixed integer programs. To reduce the number of binary variables, changes can be made on the model formulation in the same way as for EQUP-NEW-(1). Nevertheless, owing to the great amount of computation required for even small problems, a heuristic solution procedure based on Jacobsen's OSDP procedure (One Shot Dynamic Programming) for the dynamic plant location problem is proposed.[238]

The dynamic equipment optimization model (dynamic plant location model) can be approximately described as a shortest path problem. However, this approach can only be proved to yield an optimal solution if there is just one type of equipment (location) considered. For equipment selection (multilocation problems), where several types of equipment (locations) are given, it might produce a heuristic solution of reasonable quality. Thus dynamic programming is used with the heuristic assumption that the configuration (state) optimal for arrival at a period (stage) is also optimal for departure from that period (stage). Without this restrictive assumption the optimal solution can only be found if, in addition, suboptimal configurations at the arrival of a period (stage) are examined for expansion, as they still might be on the optimal expansion path for the *complete* time horizon. Using the restrictive assumption, however, at each period (stage) only one configuration (state) is examined as a basis for expansion and the 'curse of dimensionality' is circumvented (see alg.8.6). For this reason Jacobsen named his procedure one shot dynamic programming.[239]

---

238   Jacobsen, S.K.: Heuristic solution to dynamic plant location problems, in: M. Roubens (Ed.), Advances in Operation Research, Amsterdam 1977, pp.207-211

239   Jacobsen, S.K.: Heuristic solution to dynamic plant location problems, in: M. Roubens (Ed.), Advances in Operation Research, Amsterdam 1977, p.209

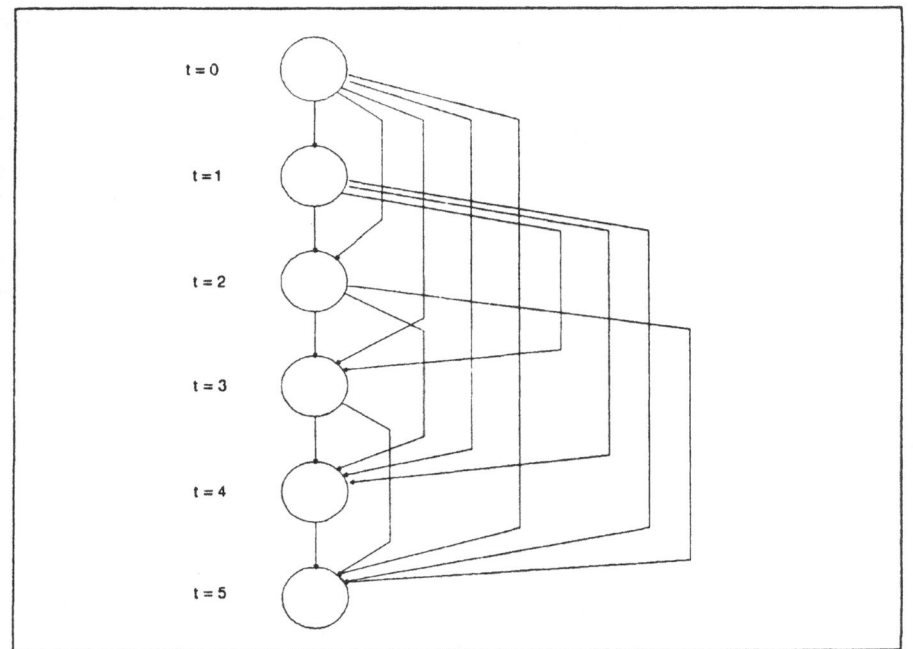

fig. 8.6: OSDP procedure [240]

In the graph (see fig. 8.6) each node t corresponds to a time period and the arc length $d(t,t+\theta)$ to an expansion step from one period t to another following period $t+\theta$. Thus the arc length in the dynamic equipment optimization model is given by:

Objective function:

$$d(t,t+\theta) = \min \sum_{m=1}^{M} \sum_{b=1}^{B_m} F_{bmt} \cdot z_{bmt} + \sum_{s=t}^{t+\theta} \sum_{g=1}^{G} \sum_{r=1}^{R_g} CF_{grs} \cdot x_{grs} \quad (8.3.1.4.5)$$

Constraints:

$$\sum_{g=1}^{C} \sum_{r=1}^{R_g} x_{grs} \cdot W_{mgr} \leq \sum_{u=0}^{t} \sum_{b=1}^{B_m} b \cdot z_{bmu} \quad \forall\, m,s \quad \text{with } s=\{t,\ldots,t+\theta\} \quad (8.3.1.4.6)$$

$$\sum_{r=1}^{R_g} x_{grs} = frac_{gs} \cdot R_{smin} \quad \forall\, g,s \quad \text{with } s=\{t,\ldots,t+\theta\} \quad (8.3.1.4.7)$$

$$x_{grs} \geq 0 \quad \forall\, s,g,r \qquad z_{bms} = \{0,1\} \quad \forall\, s,m,b \quad \text{with } s=\{t,\ldots,t+\theta\} \quad (8.3.1.4.8)$$

Note that the formulation of the arc length problem is almost indentical to the single-period equipment optimization model EQUP-NEW-(1), except for the fact that now production over several periods, i.e. over $\theta + 1$ periods, is considered. Thus with slight changes the solution procedures of model EQUP-NEW-(1) can be used.

Furthermore, if $T + 1$ is the total number of time periods considered, the graph consists of $T \cdot (T + 1)/2$ arcs. This can be very cumbersome to solve, as for each arc the single-period formulation, i.e. an adapted form of model EQUP-NEW-(1), has to be evaluated. Thus it is suggested that an adapted version of the greedy heuristic in chapter 8.3.1.3 be used to calculate a heuristic solution for each arc (alg.8.7).[241]

Below linear payment functions are assumed for the servers at each cell or the transportation system. In the first period they consist of variable payments FV_m for each machine, vehicle etc. and fixed payments for the cell or transportation system FF_m:

```
FF_m + FV_m·s_m

  |        |
  |        L number of servers
  |     L payments for each machine, vehicle of type m during the first period
  L payments for each cell, transportation system of type m during the first period
```

Payments for equipment expansion in later periods $t > 1$ have the following form:

```
FFE_m + FVE_m·s_m

  |        |
  |        L number of servers
  |     L payments for each machine, vehicle of type m in periods with t>1
  L payments for adding a cell, transportation system of type m in periods with t>1
```

However, any increasing function for equipment payments can be used. Note that these payments still have to be discounted to obtain their present value.

```
┌─────────────────────────────────────────────────────────────────────────────┐
│ Arc generation                                                                │
├─────────────────────────────────────────────────────────────────────────────┤
│                                                                               │
│ for t=0 to T                                                                  │
│     set time horizon: hor=t                                                   │
│     for n=0 to hor                                                            │
│         set period beginning: pa=hor-n                                        │
│         if pa > 0 then set starting configuration vector NS                   │
│          equal to the best configuration for pa-1;                            │
│         perform subroutine EQSUB(NS,pa,hor,cost)                              │
│         update cheapest expansion path p_t for time horizon hor               │
│ stop                                                                          │
│                                                                               │
└─────────────────────────────────────────────────────────────────────────────┘
```

alg. 8.6: Algorithm for arc generation

---

subroutine EQSUB($\underline{NS}$,pa,hor,cost)

Step 0:

initialize $SMW_{mt} = 0 \quad \forall m, \ pa \le t \le hor$

$SR_{gt} = frac_{gt} \cdot R_{tmin} \quad \forall g, \ pa \le t \le hor$

$MAXSMW_m = NS_m$

$sx_{grt} = 0 \quad \forall g,r$

$Q = \dfrac{1}{(1+i)^{pa}}$

Step 1:

find $\min\limits_{\substack{g,r \\ SR_{gt} > 0}} \left[\sum\limits_{m \in R_g} (CO_{mgr} \cdot Q + \sigma_m + \beta_m) \right]$

with $\sigma_m = \begin{cases} (FFE_m \cdot Q) \cdot W_{mgr}/PMIN & \text{if } MAXSMW_m = 0 \text{ and } pa > 1 \\[4pt] FF_m \cdot W_{mgr}/PMIN & \text{if } MAXSMW_m = 0 \text{ and } pa = 1 \\[4pt] 0 & \text{if } MAXSMW_m > 0 \end{cases}$

and $\beta_m = \begin{cases} FVE_m \cdot Q \cdot W_{mgr}/PMIN & \text{if } MAXSMW_m \in N_0 \text{ and } pa > 1 \\ & \text{and } \max\limits_{t} \{SMW_{mt}\} = (MAXSMW_m \cdot EF_m) \\[6pt] FV_m \cdot W_{mgr}/PMIN & \text{if } MAXSMW_m \in N_0 \text{ and } pa = 1 \\ & \text{and } \max\limits_{t} \{SMW_{mt}\} = (MAXSMW_m \cdot EF_m) \\[6pt] 0 & \text{if } MAXSMW_m \text{ noninteger} \\ & \text{or } \max\limits_{t} \{SMW_{mt}\} < (MAXSMW_m \cdot EF_m) \end{cases}$

Step 2:

find $\min\limits_{x_{g^*r^*}} \{ x_{g^*r^*} \cdot W_{mg^*r^*} = ([MAXSMW_m] - MAXSMW_m) \cdot EF_m \} \quad \forall m$ [242]

calculate $\delta SR_{g^*t} = SR_{g^*t} - x_{g^*r^*} \quad \forall g, \ pa \le t \le hor$

---

242 The variable in square brackets [ ] is rounded to the next higher integer value.

Step 3:

for all t with pa≤t≤hor

   if $\delta SR_{g^*t} \geq 0$

     set $SMW_{mt} = SMW_{mt} + x_{g^*r^*} \cdot W_{mg^*r^*}$ ∀ m , pa≤t≤hor

             $sx_{g^*r^*t} = sx_{g^*r^*t} + x_{g^*r^*}$ pa≤t≤hor

             $SR_{g^*t} = SR_{g^*t} - x_{g^*r^*}$ pa≤t≤hor

   if $SR_{g^*t} > 0$ and $\delta SR_{g^*t} < 0$

     set $SMW_{tm} = SMW_{tm} + SR_{g^*t} \cdot W_{mg^*r^*}$ ∀ m, pa≤t≤hor

             $sx_{g^*r^*t} = sx_{g^*r^*t} + SR_{g^*t}$ pa≤t≤hor

             $SR_{g^*t} = 0$ pa≤t≤hor

Step 4:

if $MAXSMW_m \leq \max_{t} \{SMW_{mt}/EF_m\}$ then $MAXSMW_m = \max_{t} \{SMW_{mt}/EF_m\}$ ∀ m , pa≤t≤hor

Step 5:

if $SR_{gt} = 0$ ∀ g, pa≤t≤hor

   then stop: sx_{gr} equals the production flow of each route

   $MAXSMW_m$ rounded to the next higher integer value yields the number of

   machines/set-up tables/vehicles of type m: $s_m = [MAXSMW_m]$;

else go to step 1.

alg. 8.7: Algorithm for arc length calculation

Example

Use of the above model is demonstrated in the following example. It is based on the input data of the examples in chapter 8.1.2.3 (tab.8.1) and chapter 8.1.2.4 (tab.8.5). It consists again of a part family of 3 different part types {part1, part2, part3}, where each part type forms a part group. For each part type different, alternative routes exist, i.e. part1 and part2 have two, and part3 has three alternative routes. The input data on production requirements for each part type for every single period is given in tab.8.15. It is assumed that the systems lifetime PMIN comprises 5 periods.

	part1	part2	part3
R_{g0}	6000	8000	7000
R_{g1}	6000	8000	8000
R_{g2}	8000	8000	9000
R_{g3}	11000	16000	10000
R_{g4}	9000	15000	11000
R_{avg}	8000	11000	9000

tab. 8.15: Production requirements

Linear payment functions are assumed for the servers at each cell or the transportation system. Their values are given in tab.8.16. The rate of return used has the value of 15%.

	L/U St.	mill-1	mill-2	mill-3	vtl-1	vtl-2	transp.
FF_m	10000	50000	100000	100000	100000	50000	500000
FV_m	50000	1000000	800000	1200000	1500000	800000	20000
FFE_m	20000	70000	120000	150000	150000	100000	1000000
FVE_m	50000	1100000	850000	1300000	1600000	850000	20000
EF_m	0.85	0.85	0.85	0.85	0.85	0.85	0.50

tab. 8.16: Payments for equipment expansion

From the above heuristic the following configuration (tab.8.17) and part flows (tab.8.18) can be obtained. Note that at the beginning of period 3 a system extension takes place.

	L/U St.	mill-1	mill-2	mill-3	vtl-1	vtl-2	transp.
s_{0m}	2	1	2	2	1	1	2
s_{1m}	2	1	2	2	1	1	2
s_{2m}	2	2	3	3	2	1	3
s_{3m}	2	2	3	3	2	1	3
s_{4m}	2	2	3	3	2	1	3

tab. 8.17: Configuration of the resulting flexible manufacturing system

	part1		part2		part3		
	route1	route2	route1	route2	route1	route2	route3
x_{gr0}	4889	1111	8000	0	1630	5370	0
x_{gr1}	4889	1111	8000	0	1630	6370	0
x_{gr2}	8000	0	0	8000	0	9000	0
x_{gr3}	11000	0	7000	9000	0	10000	0
x_{gr4}	9000	0	6000	9000	0	11000	0

tab. 8.18: Production per period on each route

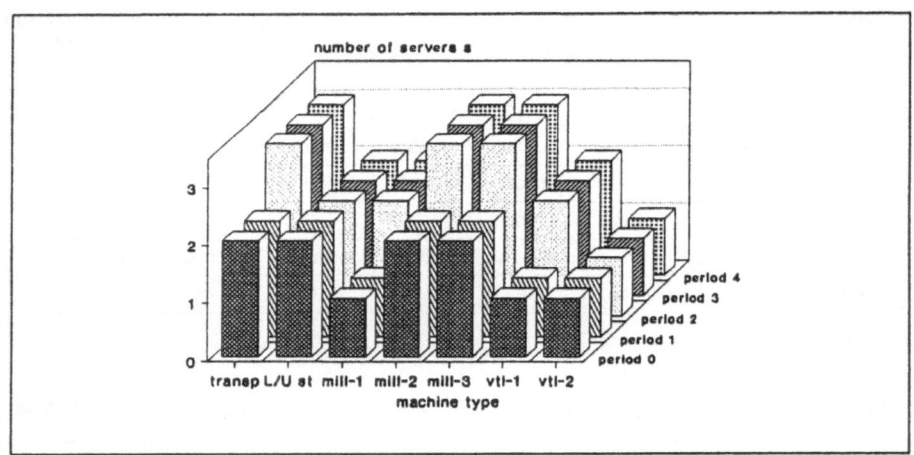

fig. 8.7: Configurational change over the system's lifetime

Discussion

Input:
- Different alternative routes must be provided.
- Production requirements for each part group and all production periods must be available.

Model:
- Because of static system behavior modelling the model is restricted to flexible transfer lines and multi-lines or to flexible manufacturing systems with a large number of pallets, where costs dependent on the number of pallets are negligible.
- The model makes allowance for flexibility aspects. This implies that changing levels in production requirements during the existence of a system are anticipated.

- The cell structure of a flexible manufacturing system is considered in the cost model.
- Possible cost reductions through the performance of a certain combination of operations are neglected.
- Production criteria other than satisfying demand - e.g. throughput time - cannot be included.
- Possible machine breakdowns are not explicitly included in the model. However, an efficiency factor can be used to reflect approximately the expected annual downtime for a cell.
- Pallet, fixture and in-process inventory costs are neglected.

Procedure:
- For practical problem sizes only a heuristic solution procedure is applicable.

8.3.2 Models with a limited number of pallets

Models with a limited number of pallets incorporate queueing theory to allow the consideration of dynamic system behavior. When equipment selection occurs, this kind of modelling is especially applicable to flexible machining systems, owing to the stochastic nature of the part flow in such a system. Furthermore, if the costs dependent on the number of pallets in the system are significant, the number of pallets should be included in the selecting process.

8.3.2.1 A new model for optimal equipment selection with maximum part flow

This new model for optimal equipment selection incorporates dynamic system behavior. Different pieces of equipment are selected according to the objective of minimum equipment costs and in-process inventory costs. Operating costs are neglected. The flexible manufacturing system is configured so as to allow a given minimum production rate R_{min} of a given part family to be attained.

Model EQLP-NEW-(1)

Indices:
b	:	index for number of servers (machines, set-up tables, carriers etc.) at cell
g	:	part group index
m	:	index for cells or transportation system
r	:	index for routes

Decision variables:
N	:	number of pallets in the system
q_{gr}	:	flow fraction of part group g on route r
z_{bm}	:	binary integer variable which is one if cell (transportation system) m with the amount of b machines (vehicles) is included in the configuration of the flexible manufacturing system.

Parameters:

CC_g	:	capital costs for a part of part group g
$frac_g$	:	product rate, i.e. the fraction part group g has of total production
I_{bm}	:	equipment costs for cells or transportation systems having b machines or respectively b vehicles
IP	:	pallet costs
IFI_g	:	fixture costs for parts of part group g
N	:	number of pallets
N^b	:	lower bound for the number of pallets
N_{max}	:	maximal number of pallets
N_+	:	ABA-bound estimate of N
R_{min}	:	required production rate for the system
SW_{min}	:	minimum workload of the system
$T(N,\underline{z},g)$	:	throughput rate of the system
W_{gmr}	:	workload of part group g at cell m on route r

Objective function:

$$\min \sum_{m=1}^{M} \sum_{b=1}^{B_m} I_{bm} \cdot z_{bm} \;+\; IP \cdot N \;+\; N \cdot \sum_{g=1}^{G} (IFI_g + CC_g) \cdot frac_g \qquad (8.3.2.1.1)$$

$$\underbrace{\phantom{\sum_{m=1}^{M} \sum_{b=1}^{B_m} I_{bm} \cdot z_{bm}}}_{\substack{\text{cell or} \\ \text{transportation} \\ \text{system costs}}} \qquad \underbrace{}_{\substack{\text{pallet costs, fixture costs} \\ \text{and in-process inventory costs}}}$$

Constraints:

$$\sum_{r=1}^{R_g} q_{gr} = frac_g \qquad \forall\ g,\ \text{with}\ \sum_{g=1}^{G} frac_g = 1 \qquad (8.3.2.1.2)$$

$\llcorner$ product rate of part group g
$\llcorner$ fraction of part group g which is produced on route r

$$T(N,\underline{z},g) \geq R_{min} \qquad (8.3.2.1.3)$$

$\llcorner$ minimum required production rate
$\llcorner$ throughput rate of the system

$$N \leq N_{max} \qquad (8.3.2.1.4)$$

$\llcorner$ maximal number of pallets
$\llcorner$ number of pallets in the system

$$0 \leq q_{gr} \leq 1 \qquad (8.3.2.1.5)$$

$$z_{bm} \in \{0,1\},\ N \in Z^+ \qquad (8.3.2.1.6)$$

In the objective function the annual equipment costs for cells and the transportation system are considered first. The binary decision variable z_{bm} decides whether b machines, set-up tables or vehicles of type m causing I_{bm} costs are included in the system. Hereby it is assumed, that I_{bm} is increasing with higher b.

Next the annual pallet costs IP for one pallet are multiplied by the number of pallets N in the system to yield total pallet costs.

To obtain annual in-process inventory costs, first the capital costs CC_g for a single part of part group g is multiplied by the fraction $frac_g$ each part group has of total production. The summation over all part groups can be considered as the average annual inventory costs for one part. Multiplying it by the average number of pallets N which are constantly in the system yields the annual in-process inventory costs. Fixture costs IFI_g are obtained in an analogous way .

In the first constraint set (8.3.2.1.2) the summation of all fractions q_{gr} over all routes of part group g must be equal to the fraction $frac_g$ part group g has of total production.

Next the throughput rate $T(N,\underline{z},\underline{q})$ is equal to or greater than a required production rate R_{min} (8.3.2.1.3), and in (8.3.2.1.4) the number of pallets is less than or equal to the maximal number of pallets N_{max} allowed in the system.

Finally the decision variables are defined in the remaining constraint sets.

Procedure for solution

The suggested procedure for solution is to use implicit enumeration. However, before the procedure is described, some helpful properties of the model formulation in conjunction with the throughput function are stated.

In chapter 7.3. it was already proved that a minimal cost solution obtained by static system behavior modelling yields a lower cost bound configuration on server costs.

An alternative explanation for this fact can be given by the following considerations: The above model formulation can be seen as an extension of the model CALP-Vinod/Solberg. The part flow is now included in the decision process instead of being given as input data. According to the solution procedure provided by Dallery and Frein, a lower bound vector must be generated. However, dependent on the part flow selected, several combinations of ABA-bounds are possible in this new model.[243] Of these bounds, the one with the lowest server costs must be chosen. Thus the static model of EQLP-NEW-(1), i.e. model EQUP-NEW-(1) with operating cost equal to zero can be interpreted as a generator to find the least cost ABA-bound.[244]

Theorem 6: The ∞-monotonicity property holds for the throughput function of closed queueing networks with flow-optimal workloads W^* for each number of pallets N:

$$T(N,W^*) \leq T(N+1,W^*) \qquad \forall\ N$$

Proof: For constant workloads this was already proved by Suri.[245] Hence, if at each increasing step of the number of pallets, the workload is changed in such a way that the throughput is improved, Theorem 6 must hold. ∎

Conjecture 7: The throughput function of closed queueing networks with multiple-server stations and optimal workloads W^* is concave in the number of pallets N.

---

243   For a detailed description on ABA-bounds see for example Zahorjan, J., Sevcik, K.C., Eager, D.L., Galler, B.: Balanced Job Bound Analysis of Queueing Networks, in: Comm. ACM, 25(1982)2, pp.134-141

244   see chapter 8.3.1.3

245   Suri, R.: A Concept of Monotonicity and Its Characterization for Closed Queueing Networks, in: OR, 33(1985)3, pp.606-624; See also chapter 6.1.2.2.3 for an overview on the properties of the throughput function.

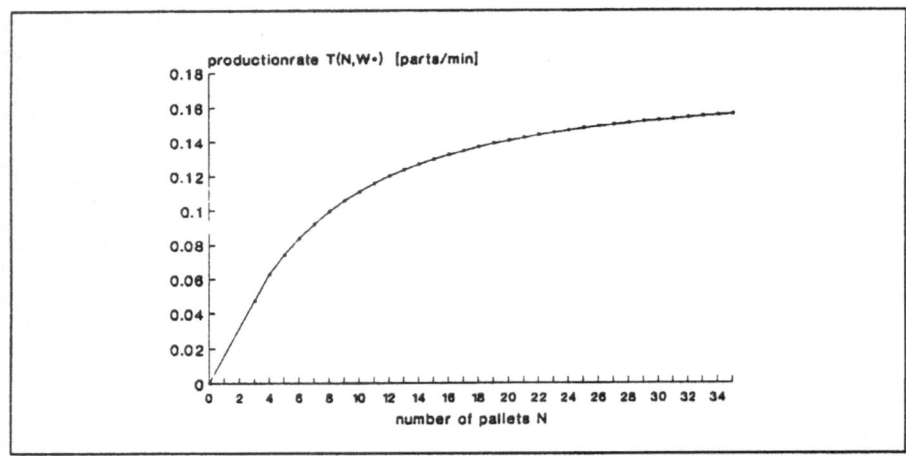

fig. 8.8: Throughput function $T(N, W^*)$

For constant workloads W this was proved by Shanthikumar and Yao.[246]

Based on Conjecture 7 a lower bound for the number of pallets N^b necessary to reach a given production rate R_{min} can be provided:

$$N^b = \frac{R_{min} - T(W^*,N)}{T(W^*,N) - T(W^*,N-1)} + N \leq N^*$$

It is derived by dividing the remaining gap between the actual and the required production rate $(R_{min} - T(W^*,N))$ by the increase in the production rate caused by adding the last pallet. Based on the assumption of concavity stated in Conjecture 7, this yields a lower bound estimate on the number of pallets still necessary to achieve the required production rate R_{min}. It should also be noted, that owing to concavity the difference in the production rate by adding a new pallet, i.e. $T(W^*,N) - T(W^*,N-1)$, decreases with increasing N, thus the bound gets tighter with increasing N.

The solution procedure can be outlined as follows:

Based on the result of Theorem 1, model EQUP-NEW-(1) is solved to provide a static lower cost bound for the servers $\underline{z}_+$ in the first step. Furthermore a lower ABA-bound on the number of pallets N_+ is evaluated.

Afterwards, in step 2 + 3, a feasible solution is generated. Based on the lower cost bound vector $\underline{z}_+$, the number of pallets is increased in steps starting from N_+. At each step the throughput of the system is maximized with the help of model ROLP-NEW-(1). When the required production rate is reached, it is known from Theorem 6, that this must be the cost optimal number of pallets N^* for the given server vector $\underline{z}$. However, it is possible that the maximal number of pallets is exceeded before the required throughput is obtained, or that the lower bound of the number of pallets N^b is larger than N_{max}. Then the bottleneck

246   Shanthikumar, J.G., Yao, D.: Second-Order Properties of the Throughput of a Closed Queueing Network, in: Mathematics of OR, 13(1988)3, p.525; See also chapter 6.1.2.2.3 for an overview on the properties of the throughput function.

cell is increased by another server, the number of pallets reset, and again the throughput is maximized while the number of pallets is increased in steps. This is repeated until the first feasible solution is found.

In step 4 an improvement of the feasible solution is tried, when the configuration vector z of the feasible solution differs from the static solution z_+. This is done by a simple exchange procedure, i.e. from each type of machine, vehicle etc. one server is subtracted (if possible) and added successively to each of the other types. Each time, the optimal flow q and number of pallets N is evaluated in the same way as before for the feasible solution, i.e. the number of pallets is increased in steps and the throughput maximized. Moreover at each step it is checked whether system costs with the lower bound number of pallets (z, N^b) are larger then system costs for the best solution found so far, or whether the lower bound number of pallets N^b is larger than N_{max}. If one (or both) of these conditions is (are) fulfilled, the configuration is discarded. Otherwise the number of pallets is increased and throughput maximized until either one (or both) of these conditions is (are) fulfilled, or the minimal production rate R_{min} is achieved and thus an improved feasible solution is obtained.

In the last step an implicit enumeration is performed. Starting from the $\underline{0}$-vector, all possible combinations of server configurations z are checked. A lower cost bound on the servers is provided with the equipment costs of z_+ from step 1, an upper cost bound on system costs by the feasible solution of step $2+3+4$. In addition a feasibility check is made, whether all parts can be produced on the current configuration. When all these checks have been completed, the procedure continues in the same manner as in step four and evaluates the optimal flow vector q and the number of pallets N. At the end of the procedure, the best solution found is the optimal solution.

Implicit enumeration for EQLP-NEW-(1)

Step 1:

solve model EQUP-NEW-(1) with $CO_{gr}=0 \ \forall \ g,r$ to obtain static lower cost bound SLB with the server configuration $\underline{z}_+$

set $\underline{z} = \underline{z}_+$, $N_+ = SW_{min} \cdot R_{min}$

Step 2:

set $N=N_+ -1$

Increase N in steps and evaluate each time the maximal throughput flow q with ROLP-NEW-(1) for $(\underline{z}, N)$ until

$- \ T(W^M, N) \geq R_{min}$

$- \ or \ N_{max} < N^b$

with $N^b = \dfrac{R_{min} - T(W^M, N)}{T(W^M, N) - T(W^M, N-1)} + N$

Step 3:

if $T(W^M, N) \geq R_{min}$

    then the best solution known so far is the feasible solution given by $BS=(\underline{z}, N, \underline{q})$

else add a server to the bottleneck cell, go to Step 2.

Step: 4

if $\underline{z}_+ \neq \underline{z}$ improve the feasible solution by generating alternative

configurations $\underline{z}$ by exchanging servers between different stations $m=1,..M$ in

subtracting one server from one station m^M (if available) and add it

successively to one of the other stations $m \neq m^M$

    for each configuration vector $\underline{z}$

        set $N=N_+ -1$

        increase N in steps and evaluate each time the maximal throughput flow $\underline{q}$ with ROLP-NEW-(1) for $(\underline{z}, N)$ until

        - $T(W^M, N) \geq R_{min}$

        - or $N_{max} < N^b$

        - or equipment costs of $(\underline{z}, N^b) \geq BS$

        if $T(W^M, N) \geq R_{min}$ and equipment costs of $(\underline{z}, N) < BS$

            then an improved solution is given by $BS=(\underline{z}, N, \underline{q})$

        else discard $\underline{z}$

```
Step 5:

for all possible configurations enumerate z starting with z=0:

    if for a given z

      - server costs are less then SLB
      - server costs are more then system costs of BS
      - production of a product group is impossible

        discard z

    else

        set N=N₊-1

        increase N in steps and evaluate each time the maximal throughput flow g
        with ROLP-NEW-(1) for (z,N) until

        - T(W*,N) ≥ Rₘᵢₙ

        - or Nₘₐₓ <Nᵇ

        - or equipment costs of (z,Nᵇ) ≥ BS

        if T(W*,N) ≥ Rₘᵢₙ and equipment costs of (z,N) < BS

            then an improved solution is given by BS=(z,N,g)

        else discard z

stop: best solution found BS is optimal solution
```

alg. 8.8: Implicit enumeration for EQLP-NEW-(1)

Example

Use of the above model is demonstrated below. For the input data, i.e. part data, refer to chapter 8.1.2.3 (tab.8.1). Each part type forms a single part group. In tab.8.19 the configuration obtained through static system behavior modelling (stat.) from model EQUP-NEW-(1), the feasible solution (feas.) from step 3, and the optimal configuration with minimal equipment costs is given. Note that for convenience

$$s_m = \sum_b b \cdot z_{bm}.$$

	L/U St.	mill-1	mill-2	mill-3	vtl-1	vtl-2	transp.	N
s_m(stat.)	2	2	2	1	1	1	2	-
s_m(feas.)	2	2	2	1	1	1	2	26
s_m(opt.)	2	2	2	1	1	1	3	22

tab. 8.19: Obtained configurations for the flexible manufacturing system

	part1	part2	part3
$frac_g$	0.370	0.333	0.296
R_{min}	10000	9000	8000
IP_g	40000	60000	50000

tab. 8.20: Production requirements and pallet costs

In tab.8.20 the production requirements and pallet costs are given. For simplicity fixture costs and inventory costs for each part type are assumed to be zero. The optimal part flow, i.e. the flow fractions q_{gr} and the production rate T_{gr} obtained for the feasible solution and the optimal solution are given in tab.8.21.

	part1		part2		part3		
	route1	route2	route1	route2	route1	route2	route3
q_{gr} (feas.)	0.342	0.029	0.309	0.024	0.296	–	–
T_{gr} (feas.)	9224	779	8348	652	8002	–	–
q_{gr} (opt.)	0.341	0.030	0.317	0.017	0.296	–	–
T_{gr} (opt.)	9270	805	8612	454	8060	–	–

tab. 8.21: Flow fraction on each route

Tab.8.22 shows the equipment costs for the feasible and the enumerated solution.

	equipment costs	system costs/per.
feas. sol.	9,440,354	5,938,812
opt. sol.	9,261,838	5,850,665

tab. 8.22: Costs for equipment

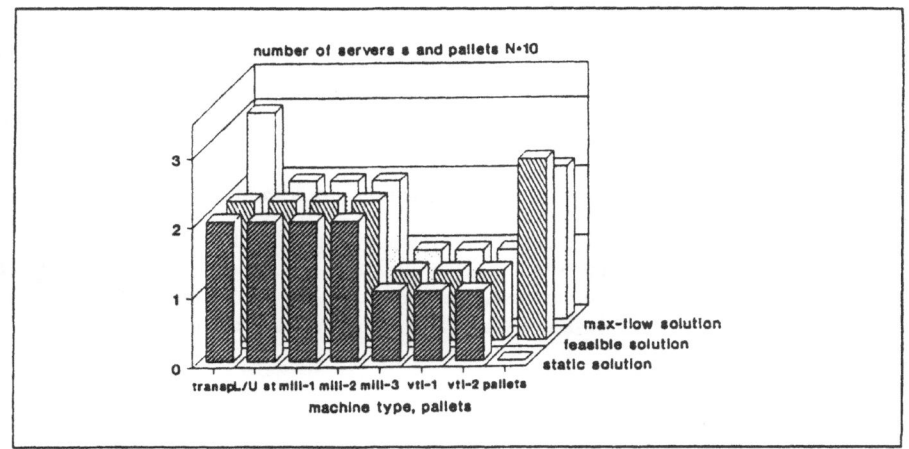

fig. 8.9: Comparison of different configurations

Discussion

Input:
- Different alternative routes must be provided.

Model:
- Dynamic system behavior modelling allows the application of this model to flexible machining systems.
- The cell structure of a flexible manufacturing system is considered in the cost model.
- Operating costs are neglected, even though they might play an important role during the design process if they have a considerable influence on costs for the flexible manufacturing system.
- Only a single period is considered. This implies that changing levels in production requirements during the existence of such a flexible manufacturing system are not anticipated. However the model can be applied several times on the static configurations obtained from a multiperiod model with unlimited pallets such as the model EQUP-NEW-(2). Then each configuration vector obtained from EQUP-NEW-(2) for one or several periods is used as a starting vector to generate a feasible solution. To allow acceptable computational requirements, first the one shot dynamic programming with the greedy heuristic can be used to solve EQUP-NEW-(2) heuristically, followed by the above described procedure for EQLP-NEW-(1) performed from step 2-4 for each configuration vector from EQUP-NEW-(2).
- The model is limited to one universal pallet type. A system with different, specialized pallets yielding different system behavior cannot be considered.
- Further limitations are given by the stochastic assumptions of closed queueing network

theory.[247]

Procedure:
-   Under the assumption that the above conjecture holds, an optimal solution is obtained.

8.3.2.2 Extension: considering operating costs during equipment optimization

The equipment optimization model in the previous chapter led to a configuration of the flexible manufacturing system such that a given minimum production rate R_{min} of a given part family is reached, while system costs are minimized. However, operating costs were neglected. Below an extension is shown which also incorporates operating costs.

Model EQLP-NEW-(2)

Indices:
b	:	index for number of servers (machines, set-up tables, carriers etc.) at cell
g	:	part group index
m	:	index for cells or transportation system
r	:	index for routes

Decision variables:
N	:	number of pallets in the system
q_{gr}	:	flow fraction of part group g on route r
z_{bm}	:	binary integer variable which is one if cell (transportation system) m with the amount of b machines (vehicles) is included in the configuration of the flexible manufacturing system.

Parameters:
CC_g	:	capital costs for a part of part group g
CO_{gr}	:	operating costs for a part of part group g
$frac_g$	:	product rate, i.e. the fraction part group g has of total production
I_{bm}	:	equipment costs for cell or transportation system having b machines or respectively b vehicles
IP	:	pallet costs
IFl_g	:	fixture costs for parts of part group g
N	:	number of pallets
N^b	:	lower bound for the number of pallets
$\bar{n}_m$	:	average queue length at cell m
N_{max}	:	maximal number of pallets
R_{min}	:	required production rate for the system
$T(N,z,g)$	:	throughput rate of the system
W_{gmr}	:	workload of part group g at cell m on route r

---

247   see chapter 6.1.2.2

Objective function:

$$\min \sum_{m=1}^{M} \sum_{b=1}^{B_m} I_{bm} \cdot z_{bm} \;+\; IP \cdot N \;+\; N \cdot \sum_{g=1}^{C} (IFI_g + CC_g) \cdot frac_g \tag{8.3.2.2.1}$$

$$\underbrace{\phantom{\sum_{m=1}^{M} \sum_{b=1}^{B_m} I_{bm} \cdot z_{bm}}}_{\substack{\text{cell or} \\ \text{transportation} \\ \text{system costs}}} \quad \underbrace{}_{\substack{\text{pallet costs, fixture costs} \\ \text{and in-process inventory costs}}}$$

$$+ \sum_{g=1}^{C} \sum_{r \in R_g(\underline{z})} \underbrace{CO_{gr} \cdot q_{gr} \cdot T(N, \underline{z}, \underline{q})}_{\text{operating costs}}$$

Constraints:

$$\sum_{r=1}^{R_g} q_{gr} = frac_g \qquad \forall\, g, \text{ with } \sum_{g=1}^{C} frac_g = 1 \tag{8.3.2.2.2}$$

where the bracket labels: q_{gr} — faction of part group g which is produced on route r; $frac_g$ — product rate of part group g

$$T(N, \underline{z}, \underline{q}) = R_{min} \tag{8.3.2.2.3}$$

where: $T(N,\underline{z},\underline{q})$ — throughput rate of the system; R_{min} — minimum required production rate

$$N \le N_{max} \tag{8.3.2.2.4}$$

where: N — number of pallets in the system; N_{max} — maximal number of pallets

$$0 \le q_{gr} \le 1 \tag{8.3.2.2.5}$$

$$z_{bm} \in \{0,1\}, \; N \in Z^{+} \tag{8.3.2.2.6}$$

The above model can be derived from the model EQLP-NEW-(1) by incorporating operating costs in the objective function. The latter are obtained by multiplying the operating costs on each route CO_{gr} by the level of production $q_{gr} \cdot T(N,\underline{z},\underline{q})$ on each route. Furthermore the constraint for the required production rate (8.3.2.2.3) is changed to an equality. Otherwise the model is identical to EQLP-NEW-(1) which is described in detail in chapter 8.3.2.1.

Procedure for solution

To solve the above model an extension of the solution procedure for model EQLP-NEW-(1) can be provided.

It is suggested that the procedure of model EQUP-NEW-(1) be applied in the same way, except that each time a feasible solution is obtained, the possibility of further reductions in costs for the whole system is investigated by applying model ROLP-NEW-(2). Thus, new pallets are added as long as the cost increase from adding another pallet is less than the cost reductions between the cost minimal routings of N and N+1 pallets in the system.

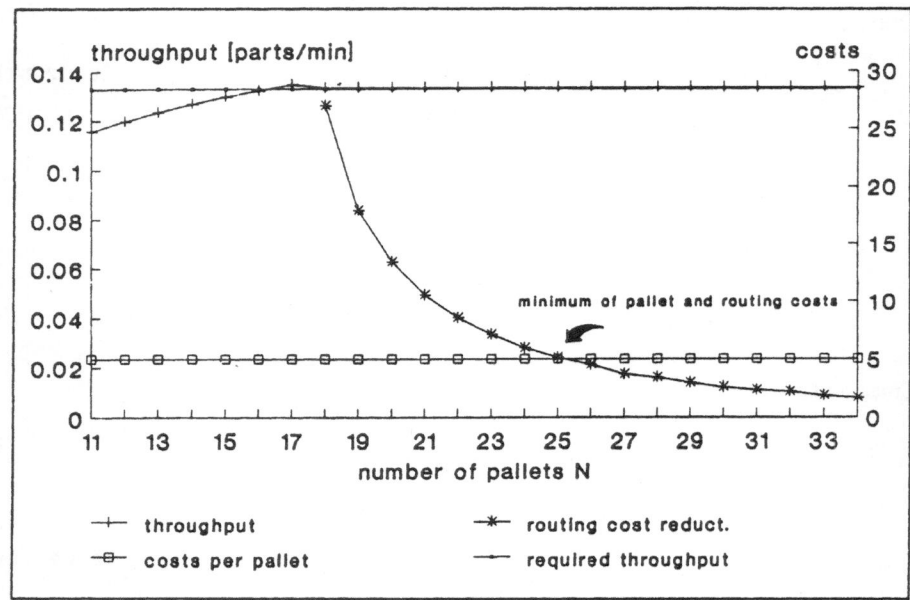

fig. 8.10: Cost savings with increasing N

To guarantee that the optimum is found, it must be shown that these cost reductions decrease with an increasing number of pallets N and cost minimal flow fractions $\underline{q}^*$ [248] (see fig.8.10). This will be demonstrated by first quantifying the cost reductions.

In the given model formulation the throughput $T(N,\underline{z},\underline{q})$ has a constant value of R_{min}. Since the throughput is an increasing concave function in the number of pallets N, the increase of throughput when the number of pallets is changed from N-1 to N

$$\Delta T = T(N,\underline{q}^*(N-1)) - T(N-1,\underline{q}^*(N-1))$$

is used to reduce operating costs and thus changes the cost minimal flow from $\underline{q}^*(N-1)$ to $\underline{q}^*(N)$:

$$\Delta T = T(N,\underline{q}^*(N-1)) - T(N,\underline{q}^*(N)) = T(N,\underline{q}^*(N-1)) - T(N-1,\underline{q}^*(N-1))$$

When flow fractions change from $\underline{q}^*(N-1)$ to $\underline{q}^*(N)$, cost changes are described by:

$$\sum_g \sum_{r \in R_g(\underline{z})} co_{gr} \cdot [q^*_{gr}(N-1) \cdot T(N-1,\underline{q}^*(N-1)) - q^*_{gr}(N) \cdot T(N,\underline{q}^*(N))]$$

Since the production rate is assumed to remain constant, i.e.

$$T(N-1,\underline{q}^*(N-1)) = T(N,\underline{q}^*(N)) = R_{min},$$

this term can be simplified to:

---

248   Cost minimal flow fractions $\underline{q}^*$ are those fractions which are cost minimal under the considerations of the given constraints.

$$\sum_g \sum_{r \in R_g(\underline{z})} CO_{gr} \cdot [q^*_{gr}(N-1) - q^*_{gr}(N)] \cdot R_{min} = \sum_g \sum_{r \in R_g(\underline{z})} CO_{gr} \cdot \Delta q^*_{gr} \cdot R_{min}$$

Small changes in Δq_{gr} can be expressed by:

$$\frac{\Delta T}{\Delta q_{gr}} \approx \frac{\partial T_N(\underline{q})}{\partial q_{gr}} = \sum_m \frac{T_N(\underline{q})}{\sum_k \sum_l t_{mkl} \cdot v_{mkl} \cdot q_{kl}} \cdot (\bar{n}_m(N-1) - \bar{n}_m(N)) \cdot t_{mgr} \cdot v_{mgr}$$

$$\Delta q_{gr} \approx \frac{\Delta T}{T_N(\underline{q})} \cdot \sum_m \frac{\sum_k \sum_l t_{mkl} \cdot v_{mkl} \cdot q_{kl}}{(\bar{n}_m(N-1)) - \bar{n}_m(N) \cdot t_{mgr} \cdot v_{mgr}}$$

Thus for the change in operating costs the following approximation is obtained:

$$\sum_g \sum_{r \in R_g(\underline{z})} CO_{gr} \cdot \Delta q^*_{gr} \cdot R_{min} \approx -\Delta T \cdot \sum_m \frac{\sum_k \sum_l t_{mkl} \cdot v_{mkl} \cdot q^*_{kl}(N)}{(\bar{n}_m(N,\underline{q}^*) - \bar{n}_m(N-1,\underline{q}^*))} \cdot \sum_g \sum_{r \in R_g(\underline{z})} \frac{CO_{gr}}{t_{mgr} \cdot v_{mgr}}$$

After approximately quantifying the cost reduction for changing numbers of pallets N and flow fractions $\underline{q}$, it now must be shown that these cost reductions decrease with an increasing number of pallets N and cost minimal flow fractions $\underline{q}^*$. For the single-server case this is demonstrated by first analyzing our approximation for an increasing number of pallets N, followed by considerations on changing flow fractions $\underline{q}^*$.

If N is increased while the flow fractions $\underline{q}$ remain constant, throughput reductions $\Delta T(N)$ are decreasing. This is because T(N) is an increasing concave function in the number of pallets N.[249]

Next the differences in queue lengths $(\bar{n}_m(N) - \bar{n}_m(N-1))$ for increasing N are examined. If it is possible to show that they remain constant or increase with increasing N, cost reductions must decrease. However, Moore has shown that if N increases asymptotically all additional pallets are added to the bottleneck cell.[250] Thus the problem arises that for the bottleneck cell the difference in queue lengths increases, whereas at other cells it must decrease, as the sum over all differences in queue lengths is equal to one, i.e.

$$\sum_m (\bar{n}_m(N) - \bar{n}_m(N-1)) = 1$$

Therefore it must be shown that because the difference in queue lengths at some cells decreases, the decrease of throughput reduction $\Delta T(N)$ has a stronger influence and thus the cost reductions converge to some finite value for large N.

For the single server case this can be achieved by using the following expression for the average queue length:[251]

$$\bar{n}_m(N) = U_m(N) + U_m(N) \cdot \bar{n}_m(N-1)$$

249 Shanthikumar, J.G., Yao, D.: Second-Order Properties of the Throughput of a Closed Queueing Network, in: Mathematics of OR, 13(1988)3, p.525; See also chapter 6.1.2.2.3 for an overview on the properties of the throughput function.

250 Moore, C.G.: Network Models for Large-Scale Time-Sharing Systems, Technical Report No.71-1, Department of Industrial Engineering, University of Michigan, Ann Arbor 1971, pp.47-52

251 Bruell, S.C., Balbo, G.: Computational Algorithms for Closed Queueing Networks, New York and Oxford 1980, p.52

The following inequality is obtained:

$$U_m(N) - U_m(N-1) \leq \bar{n}_m(N) - \bar{n}_m(N-1) \leq U_m(N)$$

By reformulating cost reductions :

$$\sum_g \sum_{r \in R_g(\underline{z})} CO_{gr} \cdot \Delta q_{gr} \cdot R_{min} \approx \sum_m \frac{-\Delta T(N) \cdot \sum_k \sum_l t_{mkl} \cdot v_{mkl} \cdot q_{kl}}{(\bar{n}_m(N) - \bar{n}_m(N-1))} \cdot \sum_g \sum_{r \in R_g(\underline{z})} \frac{CO_{gr}}{t_{mgr} \cdot v_{mgr}}$$

with

$$W_m = \sum_k \sum_l t_{mkl} \cdot v_{mkl} \cdot q_{kl}$$

$$A_m = \sum_g \sum_{r \in R_g(\underline{z})} \frac{CO_{gr}}{t_{mgr} \cdot v_{mgr}}$$

and by using the utilization law with $U_m = T \cdot W_m$ for each cell the following expression can be derived[252]

$$\sum_g \sum_{r \in R_g(\underline{z})} CO_{gr} \cdot \Delta q_{gr} \cdot R_{min} \approx \sum_m \frac{-(U_m(N) - U_m(N-1))}{(\bar{n}_m(N) - \bar{n}_m(N-1))} \cdot A_m$$

If the differences in queue lengths are approximated by the inequalities given above, the following inequalities hold for each cell:

$$\frac{-(U_m(N) - U_m(N-1)) \cdot A_m}{U_m(N)} \geq \frac{-(U_m(N) - U_m(N-1)) \cdot A_m}{(\bar{n}_m(N) - \bar{n}_m(N-1))} \geq -A_m$$

For large N the left hand side converges to zero, the right hand side has a finite value. Thus it follows that cost reductions must also converge to some finite value.

Next the question which must be examined is what happens if N is increased in steps and $\underline{q}^*$ is changed accordingly. As the process for cost reduction is started from a balanced system (maximal throughput in the single-server case is obtained by balancing the workloads[253]), cost reduction goes hand in hand with changing $\underline{q}^*$ such that the workloads are unbalanced. The more unbalanced a system is, the sooner it becomes saturated. Thus it follows that the available ΔT for cost reductions decreases even more steeply.

Based on these observations a generalization is made in Conjecture 8, which assumes that the same behavior can be observed for multi-server stations (see also empirical results in fig.8.10 and fig.8.11).

Conjecture 8: In a system consisting of single and/or multiple servers and modelled by the following minimization model for operating costs using closed queueing network theory to evaluate the production rate

252   Lazowska, E.D., Zahorjan, J., Graham, G.S., Sevcik, K.C.: Quantitative System Performance Computer System Analysis Using Queueing Network Models, Englewood Cliffs 1984, p.53

253   Shanthikumar, J.G.: On the superiority of balanced load in a flexible manufacturing system, technical report, Department of IE & OR, Syracuse University, New York, 1982, p.4

$$\min \sum_{g=1}^{C} \sum_{r \in R_g(\underline{z})} CO_{gr} \cdot q_{gr}$$

s.t. (8.3.2.2.2), (8.3.2.2.3) and (8.3.2.2.5)

costs savings from switching the production flow to cheaper routes decrease with an increasing number of pallets N.[254]

The cost reductions from switching the part flows to cheaper routes for increasing N take place until either all part groups are produced on their cheapest alternative routes or the saturation point of the system meets the required production rate, whichever comes first. The latter can for example be observed in fig.8.11, since $q_1 > 0$.

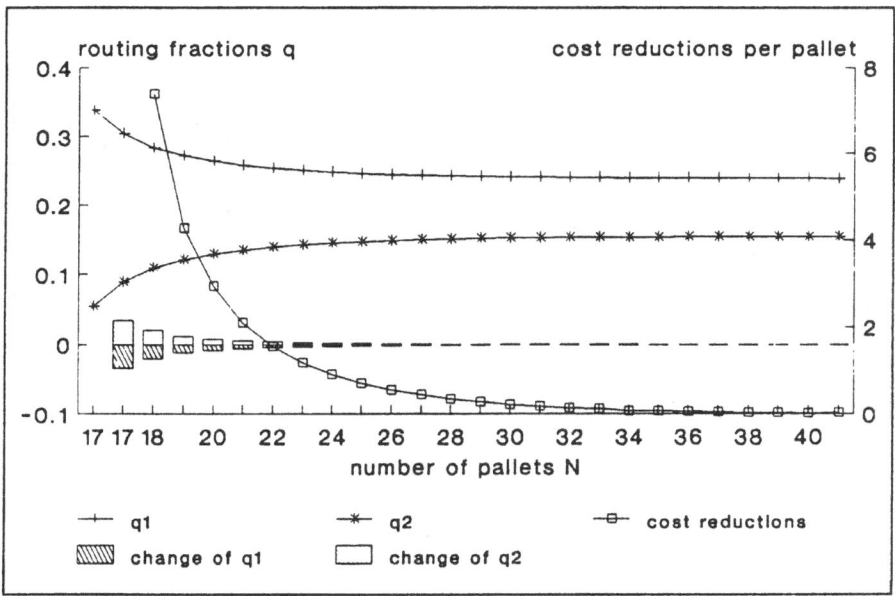

fig. 8.11: Change in routing fractions Δq_{gr} with increasing N

Implicit enumeration for EQLP-NEW-(2)
Step 1:
solve model EQUP-NEW-(1) with $CO_{gr}=0$ ∀ g,r to obtain static lower cost bound SLB with the configuration vector $\underline{z}_+$ set $\underline{z} = \underline{z}_+$, $N_+ = SW_{min} \cdot R_{min}$

---

254   Note that the objective function in the above model considers the average operating costs for one part instead of all operating costs. This simplification is due to the fact that the production rate Rmin is constant.

Step 2:

set $N=N_+-1$

Increase N in steps and evaluate each time the maximal throughput flow $\underline{g}$ with ROLP-NEW-(1) for $(\underline{z},N)$ until

- $T(W^*,N) \geq R_{min}$

- or $N_{max} < N^b$

    with $N^b = \dfrac{R_{min} - T(W^*,N)}{T(W^*,N)-T(W^*,N-1)} + N$

Step 3:

if $T(W^*,N) \geq R_{min}$

then reduce FMS costs by adding new pallets as long as

- $IP + \displaystyle\sum_{g=1}^{G} (IFI_g + CC_g)\cdot frac_g \quad < \quad \sum_{g=1}^{G} \sum_{r\in R_g(\underline{z})} CO_{gr}\cdot[q^*_{gr}(N-1) - q^*_{gr}(N)]\cdot R_{min}$

    using ROLP-NEW-(2) to evaluate the optimal $\underline{q}^*$ for each N

- and $N \leq N_{max}$

the best solution so far is the feasible solution obtained: $BS=(\underline{z},N,\underline{q})$

else add a server to the bottleneck cell, go to Step 2.

Step:4

if $\underline{z}_+ \neq \underline{z}$ improve the feasible solution by generating alternative
configurations $\underline{z}$ by exchanging servers between different stations m=1,...M in
subtracting one server from one station m^* (if available) and add it
successively to one of the other stations $m \neq m^*$

for each configuration vector $\underline{z}$

set $N=N_+-1$

increase N in steps and evaluate each time the maximal throughput flow $\underline{q}$
with ROLP-NEW-(1) for $(\underline{z},N)$ until

- $T(W^*,N) \geq R_{min}$

- or $N_{max} < N^b$

- or equipment costs of $(\underline{z},N^b) \geq BS$

if $T(W^*,N) \geq R_{min}$ and equipment costs of $(\underline{z},N) < BS$

then reduce FMS costs by adding new pallets as long as

$$- \quad IP + \sum_{g=1}^{G} (IFI_g + CC_g) \cdot frac_g < \sum_{g=1}^{G} \sum_{r \in R_g(\underline{z})} CO_{gr} \cdot [q^*_{gr}(N-1) - q^*_{gr}(N)] \cdot R_{min}$$

using ROLP-NEW-(2) to evaluate the optimal $\underline{q}^*$ for each N

- and $N \leq N_{max}$

if obtained FMS costs are less then those of BS, set BS equal to the
current solution $BS=(\underline{z},N,\underline{q})$

else discard $\underline{z}$

else discard $\underline{z}$

Step 5:

for all possible configurations enumerate $\underline{z}$ starting with $\underline{z}=\underline{0}$:

   if for a given $\underline{z}$

      - server costs are less then SLB
      - server costs are more then system costs of BS
      - production of a product group is impossible

      discard $\underline{z}$

   else

      set $N=N_+-1$

      increase N in steps and evaluate each time the maximal throughput flow $\underline{q}$ with ROLP-NEW-(1) for $(\underline{z},N)$ until

      - $T(W^{*},N) \geq R_{min}$

      - or $N_{max} < N^b$

      - or equipment costs of $(\underline{z},N^b) \geq BS$

      if $T(W^{*},N) \geq R_{min}$ and equipment costs of $(\underline{z},N) < BS$

         then reduce FMS costs by adding new pallets as long as

$$- \quad IP + \sum_{g=1}^{G} (IFl_g + CC_g)\cdot frac_g < \sum_{g=1}^{G} \sum_{r \in R_g(\underline{z})} CO_{gr}\cdot[d^{*}_{gr}(N-1) - d^{*}_{gr}(N)]\cdot R_{min}$$

           using ROLP-NEW-(2) to evaluate the optimal $\underline{d}^{*}$ for each N

         - and $N \leq N_{max}$

         if obtained FMS costs are less then those of BS, set BS equal to the current solution $BS=(\underline{z},N,\underline{q})$

         else discard $\underline{z}$

      else discard $\underline{z}$

stop: best solution found BS is optimal solution

alg. 8.9: Implicit enumeration for EQLP-NEW-(2)

Example

The following example demonstrates use of the above model. For the input data, i.e. part data, refer to chapter 8.1.2.3 (tab.8.1) and chapter 8.1.2.4 (tab.8.5). Each part type forms a part group. In tab.8.23 the configuration from static system behavior modelling (stat.) obtained through model EQUP-NEW-(1), the feasible solution (feas.), and the optimal configuration with minimal overall costs is given. Note that for convenience

$$s_m = \sum_b b\cdot z_{bm}.$$

Pallet costs and in-process inventory costs per pallet have the value of 1,000.

	L/U St.	mill-1	mill-2	mill-3	vtl-1	vtl-2	transp.	N
s_m(stat.)	2	2	2	1	1	1	2	-
s_m(feas.)	2	2	2	1	1	1	3	48
s_m(opt.)	2	2	2	1	1	1	3	48

tab. 8.23: Obtained configurations for the flexible manufacturing system

	part1	part2	part3
$frac_g$	0.367	0.333	0.300
R_{min}	11000	10000	9000
IP_g	1000	1000	1000

tab. 8.24: Production requirements and pallet costs

The optimal part flow, i.e. the flow fractions r_{gr} and the production rate T_{gr} obtained, for the feasible solution and the optimal solution is given in Tab.8.25.

	part1		part2		part3		
	route1	route2	route1	route2	route1	route2	route3
q_{gr} (feas.)	0.334	0.033	0.296	0.037	0.272	0.028	–
T_{gr} (feas.)	10017	984	8874	1125	8146	854	–
q_{gr} (opt.)	0.334	0.033	0.296	0.037	0.272	0.028	–
T_{gr} (opt.)	10017	983	8874	1125	8146	854	–

tab. 8.25: Flow fraction on each route

Tab.8.26 shows the equipment costs for the enumerated solution.

	equipment costs	FMS costs/per.
opt. sol.	8,218,000	5,460,576

tab. 8.26: Costs for equipment

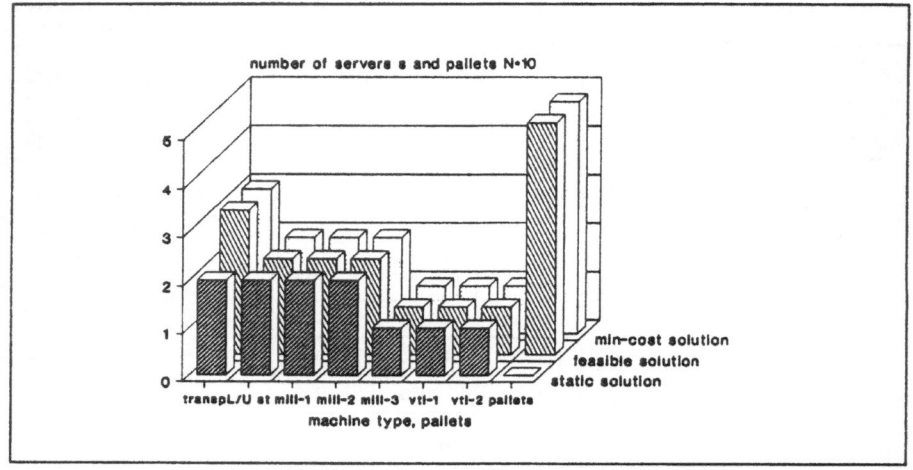

fig. 8.12: Comparison of different configurations

Discussion

Input:
- Different alternative routes must be provided.

Model:
- Dynamic system behavior modelling allows the application of this model to flexible machining systems.
- The cell structure of a flexible manufacturing system is considered in the cost model.
- Operating costs are incorporated.
- Possible cost reductions through the performance of a certain combination of operations are neglected.
- Only a single period is considered. This implies that changing levels in production requirements during the existence of such a flexible manufacturing system are not anticipated. However the model can be applied several times on the static configurations obtained from a multiperiod model with unlimited pallets like the model EQUP-NEW-(2). Then each configuration vector obtained from EQUP-NEW-(2) for one or several periods is used as a starting vector to generate a feasible solution. To allow acceptable computational requirements, first one shot dynamic programming with the greedy heuristic can be used to solve EQUP-NEW-(2) heuristically, followed by the described procedure for EQLP-NEW-(2) performed from step 2-4 for each configuration vector from EQUP-NEW-(2).
- The model is limited to one universal pallet type. A system with different, specialized pallets yielding different system behavior cannot be considered.

- Further limitations are given by the stochastic assumptions of closed queueing network theory.[255]

Procedure:
- Under the assumption that the above conjecture holds, an optimal solution is obtained.

8.4 A Model for part and equipment optimization by Whitney and Suri

The model by Whitney and Suri is a decision aid for part and machine selection. Essentially it consists of two algorithms: the PAMS (Part And Machine Selection) and the PARSE algorithm (PARt SElection). These algorithms select with different levels of detail. At the first level a rough machine and part selection is made with the help of the PAMS algorithm. Afterwards, at the second level the PARSE algorithm performs another part selection based on the previously generated configuration of the flexible manufacturing system. This is done at a more detailed level considering more restrictive constraints. Below the models on which these two algorithms are based are described in more detail.

The PAMS algorithm

The PAMS algorithm is an integer programming model designed to maximize total annual savings from parts minus annual machine tool costs.

Model PAMS

Index:
p : part type index
m : machine type index

Decision variables:
s_m : integer variable which equals the number of machine tools of type m to be included in the flexible manufacturing system.
X_p : binary variable which equals one if part p is produced on the flexible manufacturing system and equals zero if not.

Parameters:
B_p : annualized costs savings if part p is produced on a flexible manufacturing system
IV_m : annualized costs for machine tool of type m for the new system
K : annual production time
t'_{mp} : sum of annual production time of all part types p at machine tool of type m

Objective function:
$$\max_{s_m, X_p} \quad [\sum_p B_p \cdot X_p - \sum_m IV_m \cdot s_m] \tag{8.4.1}$$

---

255   see chapter 6.1.2.2

Constraints:

$$\sum_m s_m \leq S \tag{8.4.2}$$

$$\sum_p t'_{mp} \cdot X_p \leq K \cdot s_m \qquad \forall\, m \tag{8.4.3}$$

$$X_p = \{0,1\} \quad \forall\, p \quad s_m \in Z^+_0 \quad \forall\, m \tag{8.4.4}$$

In the objective function annual savings are maximized. They consist of the annualized costs B_p for producing part type p with the *current* method, which is not a flexible manufacturing system, multiplied with the decision variable X_p, which decides whether or not part type p is considered for the part family of the system. Note that Whitney and Suri give no exact specification for these savings. They refer to current production costs as costs of the current daily operations, which do not include capital recovery costs or depreciation allowances, as these were already considered at the inception of the current system. However they do not specify what is included in the production costs: "There are no universally applicable answers for all such questions."[256]

From these savings annual (amortized) costs for the new system are subtracted, i.e. amortized costs IV_m for buying machine tools of type m multiplied by the number s_m of each.

The constraint set (8.4.2) limits the number of machines selected for a flexible manufacturing system to S, whereas constraint set (8.4.3) consists of a capacity restriction. Here the sum of the annual production time t'_{mp} of all part types produced on machine type m must be less than the available capacity, consisting of annual production time K multiplied by the number s_m of machines.

Procedure for solution

The process for solving the above integer program follows a three-level procedure. Because a problem for machine and part selection can typically consists of about 5 to 20 machine tools and 50 to 5000 candidate parts, the solution procedure chosen by Whitney and Suri is heuristic.[257]

On the first level the integer constraint for the decision variables is relaxed and the model is solved as a continuous linear programming model. This is done by replacing constraint set (8.4.4) by (8.4.4'):

$$0 \leq X_p \leq 1 \quad \forall\, p \quad s_m \geq 0 \quad \forall\, m \tag{8.4.4'}$$

On the second level the machine configuration is determined. This is done through an iterative procedure, which fixes the value of one machine variable to become integer during each iterative step. The fixation is done to the machine variable closest to an integer value by introducing a new constraint:

---

256   Whitney, C.K., Suri, R.: Algorithms for Part and Machine Selection in Flexible Manufacturing Systems, in: Annals of OR, 3(1985), p.243
257   Whitney, C.K., Suri, R.: Algorithms for Part and Machine Selection in Flexible Manufacturing Systems, in: Annals of OR, 3(1985), p.252

```
if s_m < INT(s_m)  the constraint  s_m ≥ INT(s_m) is added,
if s_m > INT(s_m)  the constraint  s_m ≤ INT(s_m) is added
where INT(s_m) is the integer closest to s_m
```

Consequently the variable becomes integer during the next linear programming solution step. The procedure converges to an integer solution for the machine variables and ends when all machine variables N_1 are integer.

The part selection performed on the last level of PAMS is done in a very similar way to the machine selection. At each iterative step the part variable X_m having the largest fractional value is fixed at 1. This is continued as long as the value of the truncated solution z^n, consisting of the maximal savings obtained only by integer variables is less then savings v^k calculated by the linear programming procedure, i.e. as long as

$$v^k \; > \; \max_{1 \le n \le k} \{z^n\}$$

This indicates, that the procedure stops, when no further improvements can be obtained from fixing another part variable to one.

The PARSE algorithm

Historically the PARSE algorithm was developed to solve the loading problem for flexible manufacturing.[258] It assumes a fixed configuration of the flexible manufacturing system. In the context of part and machine selection it is used to improve the part selection obtained from the PAMS algorithm by considering resource constraints to a greater extent, i.e. tool slot constraints are considered.

Model PARSE

Index:

p	:	part type index
k	:	index for work segments of parts
m	:	index for machine types
v	:	index to distinguish individual machines within a class

Decision variables:

X_p : binary variable which has the value 1 if part p is produced on the flexible manufacturing system

x_{pkmv} : binary variable which assigns segment k of part p to an individual machine v of type m

Parameters:

$\#[A]$: number of elements of the set A

B_p : annualized costs savings if part p is produced on a flexible manufacturing system

st_{pk} : set of tools required to produce work segment k of part p

t_{pkm} : production time for work segment k of part type p at machine type m

K_m : available production time of machine type m

---

258   Whitney, C.K., Suri, R.: Algorithms for Part and Machine Selection in Flexible Manufacturing Systems, in: Annals of OR, 3(1985), p.246 and Whitney, C.K., Gaul, T.S.: Sequential Decision Procedures for Batching and Balancing in Flexible Manufacturing Systems, in: Annals of OR, 3(1985), pp.301-316

SS_m :   tool slot capacity

Objective function:

$$\max_{\{X_p, x_{pmkv}\}} \quad [\Sigma_p \, B_p \cdot X_p] \tag{8.4.5}$$

Constraints:

$$\sum_{p,k} t_{pkm} \cdot x_{pkmv} \leq K_m \quad \forall \, m, v \tag{8.4.6}$$

$$\sum_{m,v} x_{pkmv} = X_p \quad \forall \, p, k \tag{8.4.7}$$

$$\#[\,\bigcup_{p,k} \{ \sum_{p,k} st_{pk} \cdot x_{pkmv} \}\,] \leq SS_m \quad \forall \, m, v \tag{8.4.8}$$

$$X_p = \{0, 1\} \quad \forall \, p \qquad x_{pkmv} = \{0, 1\} \quad \forall \, p, k, m, v \tag{8.4.9}$$

In the objective function a maximization of annual cost savings B_p is done through part selection. Constraint set (8.4.6) ensures that the workload assigned does not exceed the available machine time. This is done in a similar way as in constraint set (8.4.3) of the PAMS algorithm. The equations in (8.4.7) describe the relationship between the variables x_{pkmv} considering operations of each part type and the part variables X_p. Restricted tool slot capacity SS_m is considered in constraint set (8.4.8). The notation #[A] denotes the number of elements of the set A. When an operation k of part type p is performed on machine v of type m the tool subset st_{pk} is needed. Then the set union of all necessary tool subsets st_{pk} is taken, to avoid counting the same tool twice. Thus the left hand side of constraint type (8.4.8) considers the number of tools necessary for all operations x_{pkmv} which are performed at each single machine v of type m. It must be less than or equal to the right hand side, given by the available tool slot capacity SS_m of a machine of type m.

Procedure for solution

The solution procedure for the PARSE algorithm is sequential and incorporates three main steps:[259]

- For each part not yet selected, evaluate a performance criterion representing the desirability of placing this part in the system.

- Select the part having the most favorable performance criterion and update tool slots used and hours consumed accordingly.

- Repeat this process until no more parts will fit in the system.

---

259   for a more detailed description of the Algorithm see: Whitney, C.K., Suri, R.: Algorithms for Part and Machine Selection in Flexible Manufacturing Systems, in: Annals of OR, 3(1985), pp.257-260 and Whitney, C.K., Gaul , T.S.: Sequential Decision Procedures for Batching and Balancing in Flexible Manufacturing Systems, in: Annals of OR, 3(1985), pp.301-316

Discussion

Input:

- The estimation of annual savings B_p for part p produced on the flexible manufacturing system seems to be a crucial point. Firstly the authors give no explanation of how these savings can be derived i.e. what kind of costs are used. Secondly a preconfigured system is already assumed before the actual optimization process for the configuration begins. This is because costs have to be considered for part p produced on a flexible manufacturing system, if a comparison with conventional systems and therefore a calculation of savings is made. These costs, however, might be dependent on the configuration of the system, which is not available at the beginning of the procedure. Thus, because the flexible manufacturing system is configured during the optimization process, only rough estimates for annual savings B_p are available ahead of this process.

Model:

- Static system behavior modelling restricts the application of this model to flexible transfer lines and multi-lines or to flexible manufacturing systems with a large number of pallets, where costs dependent on the number of pallets are negligible.

- The cell structure of a flexible manufacturing system is not considered in the cost model. Therefore cell costs are not incorporated.

- Possible cost reductions through the performance of a certain combination of operations are neglected.

- Production criteria other than satisfying demand - e.g. throughput time - cannot be included.

- Possible machine breakdowns are not explicitly included in the model. However, the production time K may be adjusted downward to reflect approximately the expected annual downtime for the machine.

- The model is only of static nature. This implies that changing levels in production requirements during the existence of a flexible manufacturing system are not anticipated.

- The question arises whether tool constraints play such an important role in the long range planning of the configuration of a flexible manufacturing system. Most systems nowadays can be equipped with powerful centralized tool supply systems, so that all necessary tools can be supplied. Therefore tool constraints often play a more important role in the operational planning system of a flexible manufacturing system, owing to time and cost savings possible when tools are more quickly available. Furthermore for flexible machining systems the actual part mix must not comprise all part types.

- Another point is the limitation of the part assignment variable X_p to a binary variable. A relaxation to allow real values between zero and one would mean that fractional production assignments between the flexible manufacturing system and conventional systems would be allowed. If this relaxation still allows to estimate the savings B_j correctly, a simplification of the solution procedure can be achieved.

- Pallet, fixture and in-process inventory costs are neglected.

Procedure:
- Only a heuristic solution is obtained.

8.5 A Model for system and equipment optimization by Sarin and Chen

The model by Sarin and Chen incorporates not only the design of a flexible manufacturing system, but also comprises the system selection between job shop, transfer line and flexible manufacturing system and the their design. The goal of the model is to find a system mix of these three types of production systems, which minimizes annual capital and operating costs. The different production capabilities of these three system designs is made comparable through an efficiency factor.[260] The part assignment to systems is performed on the aggregated level of part families. Sarin and Chen propose applying group technology techniques (e.g. cluster analysis discussed by King[261] and King and Nakornchai[262]) to group parts in families, before applying the presented model.

Model SYEQUP-Sarin/Chen

Index:

d	:	system index
i	:	part family index
m	:	machine type index
h	:	transportation system index

Decision variables:

MHS_{hd}	:	binary variable assigning the material handling system of type h to system d
x_{id}	:	binary variable, which assigns part family i to system d
s_{md}	:	integer variable signifying the number of machines of type m assigned to system d

Parameters:

BSU_{id}	:	annual batch set up costs if part family i is assigned to system d
COC_h	:	annual computer costs when material handling system m is used
EF_{md}	:	efficiency factor on machine type m when assigned to system d
IY_{id}	:	annual inventory costs
ISU_{id}	:	initial set up costs for part family i at system d
IV_m	:	annualized costs for machine type m
$MHSC_h$	:	annual material handling system costs when material handling system h is used
$MHWC_h$	:	annual labor costs per unit time of the handler
MOC_{hd}	:	annual labor costs for machine operators for material handling system h of system d
MOC_{md}	:	annual labor costs for machine operators for machine type m of system d
NS	:	number of systems
OH_{hd}	:	annual overhead costs when material handling system h is used for system d
t_{imd}	:	processing time required by part family i on machine type m when assigned to system d

260   Sarin, S.C., Chen, C.S.: A Mathematical Model for Manufacturing System Selection, in: Flexible Manufacturing Systems: Methods and Studies, Ed.: A. Kusiak, Elsevier Science Publishers B.V. (North Holland) 1986, p.102

261   King, J.R.: Machine-Component Group Formation in Group Technology, presented at the Vth Intern. Conf. on Production Research, Amsterdam Aug. 1979, in: OMEGA, 8(1980)2, pp.193-199

262   King, J.R., Nakornchai, V.: Machine-Component Group Formation in Group Technology: Review and Extension, in: IJPR, 20(1982)2, pp.117-133

Objective function:

$$\min \; \underbrace{\sum_m IV_m \cdot \sum_d s_{md}}_{\text{machine costs}} + \underbrace{\sum_m \sum_{d \in D1} MOC_{md} \cdot s_{md} + \sum_h \sum_{d \in D2} MOC_{hd} \cdot MHS_{hd}}_{\substack{\text{labor costs for a machine} \\ \text{of type m in system d}}} + \underbrace{\sum_h MHSC_h \cdot \sum_d MHS_{hd}}_{\substack{\text{material handling} \\ \text{system costs}}} \quad (8.5.1)$$

$$+ \underbrace{\sum_h MHWC_h \cdot \sum_d MHS_{hd}}_{\substack{\text{labor costs for} \\ \text{material handling}}} + \underbrace{\sum_h COC_m \cdot \sum_d MHS_{hd}}_{\substack{\text{computer system} \\ \text{costs for material} \\ \text{handling}}} + \underbrace{\sum_i \sum_d IY_{id} \cdot x_{id}}_{\substack{\text{inventory} \\ \text{costs}}} + \underbrace{\sum_i \sum_d BSU_{id} \cdot x_{id}}_{\substack{\text{batch setup} \\ \text{costs}}}$$

$$+ \underbrace{\sum_i \sum_d ISU_{id} \cdot x_{id}}_{\substack{\text{initial setup} \\ \text{costs}}} \underbrace{+ \sum_h \sum_d OH_{hd} \cdot MHS_{hd}}_{\text{overhead costs}}$$

Constraints:

$$\sum_d x_{id} = 1 \quad \forall \; i \tag{8.5.2}$$

$$\sum_i t_{imd} \cdot x_{id} \le EF_{md} \cdot s_{md} \cdot K \quad \forall \; m,d \tag{8.5.3}$$

$$\sum_m s_{md} \le \sum_h K_{hd} \cdot MHS_{hd} \quad \forall \; d \tag{8.5.4}$$

$$\sum_h MHS_{hd} \le 1 \quad \forall \; d \tag{8.5.5}$$

$$\sum_h \sum_d MHS_{hd} \le NS \tag{8.5.6}$$

$$s_{md} \in Z^+_0 \quad x_{id} = \{0,1\} \quad MHS_{hd} = \{0,1\} \tag{8.5.7}$$

The first summation of the objective function adds up the annual machine costs by taking the products of the costs IV_m for each machine with its amount s_{md} used in all systems. The annual costs of each machine tool is derived using straight line depreciation. Labor costs MOC of machine operators required for flexible manufacturing systems and transfer lines (index-set D2) are determined by the type of the material handling system MHS_{hd} used, whereas for the job shop (index-set D1) this decision is based on the number of machine tools s_{md} assigned. The costs for material handling itself can be split into three categories: annual material handling system costs $MHSC_h$, annual computer costs COC_h and annual labor costs $MHWC_h$ per unit time of the handler. Costs in conjunction with part families x_{id} are annual inventory costs IY_{id}, annual batch set-up cost BSU_{id} and initial set-up cost ISU_{id} of part family i in system d. Finally annual overhead costs OH_{hd} dependent on the material handling system used are added.

Constraints of set (8.5.2) ensure that only one part family x_{id} is assigned to one system. In the next set of constraints (8.5.3) a restriction in system capacity is made. The required processing time of machine type m in system d for part family i has to be less than the total capacity of the machine tools of type m in that system ($s_{md} \cdot K$) multiplied by an efficiency factor EF_{md}. This efficiency factor reflects the different production efficiency of each type of system (job shop, flexible manufacturing system, transfer line). Constraint set (8.5.4) ensures the assignment of a material handling system to each system. Furthermore it guarantees that the number of machines assigned do not exceed the capacity K_{hd} of the material handling system. In the last two constraint sets the number of material handling systems in a system is restricted to one. Moreover it is ensured that there are no more material handling systems than the maximal number NS of systems.

Procedure for solution

The model of Sarin and Chen is a pure integer program consisting of binary and general integer variables. To evaluate a solution the standard software program MPSX-MIP/370 was used.

Discussion:

Input:
- There is no explanation of which costs are summarized under the annual overhead costs for each system k containing the material handling system of type m.[263]

Model:
- Static system behavior modelling is applied to the job shop, the transfer line and the flexible manufacturing system or on flexible manufacturing systems with a large number of pallets, where costs dependent on the number of pallets are negligible.

- A limitation of the model is the efficiency factor. This factor has a very large influence on the choice of the production strategy used for each part family. Because this factor is given externally by the user, there are no objective criteria for its derivation.

- Another important limitation of the model is that interdependences between part and machine selection are not considered.[264] The selection of parts to form a family is largely influenced by the sequence of machines they are manufactured on, and in particular, by the type of production system which is used. If at first the parts are grouped into families, while their dependence on the production strategy is neglected, a preselection of the strategy is, so to speak, implicitly already made. This casts doubt on the value of the subsequent optimization process.

- The authors assume constant demand for each part family, whereas the demand for each part may fluctuate. Therefore no system expansion or replacements are considered.[265] Essentially this means that the workload at each machine has to remain unchanged over the whole lifetime of the production systems. This assumption is questionable, because in general the life time of a production system might be longer than the lifecycle of products they produce. It can only be true if the change in

263   Sarin, S.C., Chen, C.S.: A Mathematical Model for Manufacturing System Selection, in: Flexible Manufacturing Systems: Methods and Studies, Ed. A. Kusiak, Elsevier Science Publishers B.V. (North Holland) 1986, p.105
264   see chapter 4.2.1
265   Sarin, S.C., Chen, C.S.: A Mathematical Model for Manufacturing System Selection, in: Flexible Manufacturing Systems: Methods and Studies, Ed. A. Kusiak, Elsevier Science Publishers B.V. (North

workloads at each machine caused by parts no more produced is precisely compensated by the introduction of new parts.

- Production criteria other than satisfying demand, e.g. throughput time, cannot be included.
- Tooling costs for each part family in different production systems are not considered.
- Pallet, fixture and in-process inventory costs are neglected.
- Possible cost reductions through the performance of a certain combination of operations are neglected.

Procedure:

- The optimal solution can be evaluated by standard software.

9. Conclusions and directions for further research

At the beginning of this thesis a few introductory definitions were presented. This was followed by a description of planning problems which arise in conjunction with flexible manufacturing systems. On this basis a general planning concept for the design of production systems with an emphasis on flexible manufacturing systems was developed. This planning concept can be considered as a framework in which different tools can be applied.

It was reasoned that mathematical programming plays an especially important role in solving the design problems. Important results were obtained from the examination of relationships between static and dynamic system behavior modelling. It was stated that a flexible manufacturing system configured by static system behavior modelling has the same throughput as a system derived from dynamic modelling having an unlimited number of pallets in the system. Thus a major finding is that for a system with a large number of pallets in the system, static behavior modelling suffices for throughput considerations.

Furthermore it was shown that systems configured by static modelling yield a lower cost bound on server costs obtained from dynamic modelling. This result was significant later on when creating a solution procedure for equipment selection with dynamic system behavior modelling.

Afterwards it was demonstrated how flexibility requirements can be incorporated in the modelling process. It was argued that flexible transfer lines and multi-lines with a regular part mix, whose qualitative structure is constant, are best modelled with deterministic models, whereas a flexible machining system with an irregular part mix, whose qualitative structure fluctuates, is best modelled by stochastic models. Hence it can be concluded that static system behavior modelling is more appropriate for the design of flexible transfer lines and multi-lines, whereas dynamic system behavior modelling is applicable for the design of flexible machining systems.

Furthermore, if changing levels in production requirements for the part family of a flexible manufacturing system can be observed, these demands in long term flexibilities can be best reflected by multi-period models. Then optimization is based on the present value of cash flows, owing to possible alternative timing of the investments (e.g. investment in overcapacity at the beginning rather then an investment in system expansion later). If, however, a fixed part family over the system's lifetime is given, a single-period model is sufficient. Then the system for a given output is configured according to an average period based on cost considerations.

By classifying the models according to their decision variables and according to system behavior modelling, the existing models were positioned in a model matrix. From this matrix the major objective for further model development was obtained: new models for equipment optimization with dynamic system behavior modelling had to be developed. Two models for this class were created. Both optimize the design of a flexible manufacturing system to obtain a minimal cost configuration. Whereas the first one neglects operating costs and selects part routings so that a maximal flow of parts through the system is achieved, the second considers operating costs and generates a minimal cost flow for the obtained configuration.

Further models were also developed. A single and a multi-period model for equipment optimization were created for static system behavior modelling. Compared to existing models, these two new models allow the consideration of the cell structure of a flexible manufacturing system with its specific cost structure. Whereas the first model configures the systems according to cost data for an average period, the second is based on cash outflow considerations. This makes it possible to compare the effect of different investment timing called for in accordance with changes in the level of production required. However, an optimal solution for practical problem sizes can only be evaluated through high computational efforts. Therefore a heuristic procedure is presented. Furthermore, for the application of the model, future production requirements over the system's lifetime for each part group must be available.

Finally two models for routing optimization and one for capacity optimization were created, based on dynamic system behavior modelling. The models for routing optimization allow the distribution of production on alternative routes, so that in the first model the production rate is maximized or, in the second model, operating costs for given production requirements are minimized. Hereby convexity of the delay function and quasiconcavity of the throughput function ensure global optimality.

The new model for capacity optimization allows the determination of the optimal capacity of a system, i.e. its number of machines, vehicles, pallets etc., under a given budget constraint. Thus it is applicable in a situation where the available financial resources are restricted but not the demand on the market for the products to be produced.

An overview of modelling results, in conjunction with existing models, is given once more in fig.9.1.

system behavior → ↓ decision problems	unlimited number of pallets (static system behavior modelling) [UP]	limited number of pallets (dynamic system behavior modelling) [LP]
routing optimization [RO]	ROUP-Secco-Suardo -Kimemia/Gershwin -Avonts et al.	ROLP-Kimemia/Gershwin -Yao/Shanthikumar -NEW-(1) -NEW-(2)
capacity optimization [CA]		CALP-Vinod/Solberg -Dallery/Frein -Yao/Shanthikumar-(1) -Yao/Shanthikumar-(2) -Yao/Shanthikumar-(3) -Solot -NEW-(1)
equipment optimization [EQ]	EQUP-Graves/Whitney -Graves/Lamar -NEW-(1) -NEW-(2)	EQLP-NEW-(1) -NEW-(2)
part and equipment optimization [PAEQ]	PAEQUP-Whitney/Suri	
system and equipment optimization [SYEQ]	SYEQUP-Sarin/Chen	
part, system and equipment optimization [PASYEQ]		

fig. 9.1: Model matrix

For the planner the question arises which model to chose for his purposes. Because the answer to this question depends heavily on the individual situation, no universal recommendation can be given and the planner must tailor a planning procedure in each case by selecting appropriate tools and models according to the data available, to given restrictions and to aspects of efficiency and accuracy. The planning concept presented, the classification criteria and the discussion of limitations of each mathematical programming model in the previous chapters will give some guidance. The principal considerations about model classes available are summarized below (see fig.9.2).

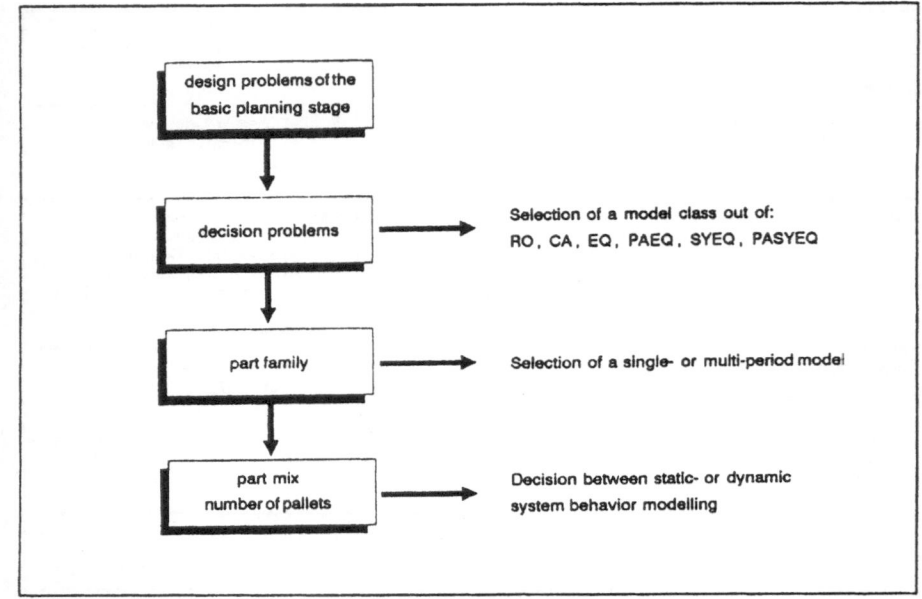

fig. 9.2: Model classes

If the model matrix is examined more closely, it can be observed that some questions are left unanswered. Hence some extensions and further improvements are desirable. In the following a list of suggestions is given:

- Development of and improvements to models are advantageous which allow part and/or system selection as well as equipment optimization.

- If not yet available, it is desirable to supplement models with a limited number of universal pallets by models for specialized pallet types.

- An extension of the models considering flexibility requirements under risk and uncertainty is desirable. So far only flexibility requirements under certainty were explicitly introduced.[266]

Appendix A:

Proof of convexity for the delay function $D_N(q)$

The proof presented in the following is basically a reformulation of the proof given by Kobayashi and Gerla using a different workload definition (see chapter 6.1.2.2.2).[267]

---

266   A sensitivity analysis to estimate the consequences of possible deviations of predicted requirements is given in Appendix C.

267   Kobayashi, H., Gerla, M.: Optimal Routing in Closed Queueing Networks, in: ACM Trans. Comp. Syst., 1(1983)4, pp.294-310;

Theorem 1: The delay function $D_N(\underline{q})$ is convex.

Proof: According to Price the following inequality for symmetric functions exits:[268]

$$\frac{C_N(\underline{a}+\underline{b})}{C_{N-1}(\underline{a}+\underline{b})} \leq \frac{C_N(\underline{a})}{C_{N-1}(\underline{a})} + \frac{C_N(\underline{b})}{C_{N-1}(\underline{b})}$$

with $C_N(\underline{x}) = \sum\limits_{\substack{i_1 + \ldots + i_m = N \\ i_j \geq 0}} x_1^{i_1} \cdot \ldots \cdot x_m^{i_m}$ and $\underline{x} = \{\underline{a},\underline{b}\}$

The $a_1,...,a_m$ and $b_1,...,b_m$ are arbitrary non-negative real numbers.

The delay function $D_N(\underline{q})$ can be derived from the throughput function $T_N(\underline{q})$ which is given by:[269]

$$T_N(\underline{q}) = \frac{C_{N-1}(\underline{q})}{C_N(\underline{q})}$$

and with the help of Little's law:[270]

$$D_N(\underline{q}) = \frac{N \cdot C_N(\underline{q})}{C_{N-1}(\underline{q})}$$

The inequality can be applied to the delay function in the following form:

$$D_N(\alpha \cdot \underline{q} + (1-\alpha) \cdot \underline{q}') = \frac{N \cdot C_N(\alpha \cdot \underline{q} + (1-\alpha) \cdot \underline{q}')}{C_{N-1}(\alpha \cdot \underline{q} + (1-\alpha) \cdot \underline{q}')}$$

$$\leq \frac{N \cdot C_N(\alpha \cdot \underline{q})}{C_{N-1}(\alpha \cdot \underline{q})} + \frac{N \cdot C_N((1-\alpha) \cdot \underline{q}')}{C_{N-1}((1-\alpha) \cdot \underline{q}')}$$

$$= \frac{N \cdot \alpha^N \cdot C_N(\underline{q})}{\alpha^{N-1} \cdot C_{N-1}(\underline{q})} + \frac{N \cdot (1-\alpha)^N \cdot C_N(\underline{q}')}{(1-\alpha)^{N-1} \cdot C_{N-1}(\underline{q}')}$$

$$= \alpha \cdot \frac{N \cdot C_N(\underline{q})}{C_{N-1}(\underline{q})} + (1-\alpha) \cdot \frac{N \cdot C_N(\underline{q}')}{C_{N-1}(\underline{q}')}$$

Thus $D_N(\underline{r})$ is a convex function:

$$D_N(\alpha \cdot \underline{q} + (1-\alpha) \cdot \underline{q}') \leq \alpha \cdot D_N(\underline{q}) + (1-\alpha) \cdot D_N(\underline{q}') \blacksquare [271]$$

---

268 Price, T.G.: Balanced Computer Systems, Stanford Electronics Lab. Technical Report No.88, Stanford University, Stanford April 1974 pp.35-49;

269 see chapter 7.2.1

270 Little, J.D.C.: A Proof of the Queueing Formula L = λ · W, in: OR 9(1961), pp.383-387

271 Another proof can be provided by the concept of strong stochastic convexity, see: Shanthikumar, J.G., Yao, D.D.: Second-Order Stochastic Properties in Queueing Systems, in: Proc. of the IEEE, 77(1989)1, p.167. The author is grateful to Dr. David D. Yao for pointing out this aspect.

Appendix B:

First order derivatives of the throughput function $T_N(\underline{q})$

By using the chain rule, the first derivative of the throughput function $T_N(\underline{c})$ can be written as:

$$\frac{\partial T_N(\underline{q})}{\partial q_{pr}} = \sum_m \frac{\partial T_N(\underline{q})}{\partial W_m} \cdot \frac{\partial W_m}{\partial q_{pr}}$$

Here the sensitivity of the average throughput function $T_N(\underline{q})$ with regard to the relative workload W_m at the (load dependent) station m is in accordance with Gordon and Dowdy:[272]

$$\frac{\partial T_N(\underline{q})}{\partial W_m} = - \frac{T_N(\underline{q})}{W_m} \cdot (\bar{n}_m(N) - \bar{n}_m(N-1))$$

The variable $\bar{n}_m$ is the average queue length at station i, if the system contains N or N-1 pallets. If we write for the relative workload W_m:

$$W_m = \sum_p \sum_r t_{mpr} \cdot v_{mpr} \cdot q_{pr}$$

its derivation with regard to q_{pr} can be stated as:

$$\frac{\partial W_m}{\partial q_{pr}} = t_{mpr} \cdot v_{mpr}$$

As the result the derivative of the throughput function $T_N(\underline{q})$ with regard to the flow fraction q_{pr} can be stated as:

$$\frac{\partial T_N(\underline{q})}{\partial q_{pr}} = \sum_m \frac{T_N(\underline{q})}{\sum_k \sum_l t_{mkl} \cdot v_{mkl} \cdot q_{kl}} \cdot (\bar{n}_m(N-1) - \bar{n}_m(N)) \cdot t_{mpr} \cdot v_{mpr}$$

Appendix C:

Consideration of demand under risk and uncertainty

In the models for equipment optimization in chapter 8.3 it is assumed that the production forecast for the system over its lifetime is accurate. This is, in many cases, an unrealistic assumption. To deal with risk and uncertainty in production demand and therefore also in flexibility demand, it is suggested that a sensitivity analysis is performed on the results of models which assume certainty. This enables the effects of forecast errors at least to be estimated. For this reason the consequences of prediction errors on the functioning of the system are examined and bounds derived which allow them to be

---

272   Gordon, K.D., Dowdy, L.W.: The Impact of Certain Parameter Estimation Errors in Queueing Network Models, in: ACM Perf. Eval. Rev., 2(May 1980) pp.3-9

estimated.

Normally, if production forecasts are correct, the sum of all production requirements per period of each part type produced on a system would yield the required average production rate T_{av}. Here the fraction each part type has of the production rate T_{av} is given by $frac_p$. The system would be configured based on these assumptions.

Changes in production volume (i.e. the production rate T_{sys}) of the system are now examined if the forecasted fraction $frac_p$ of part type p of total production changes by $\delta frac_p$.

An approximate bound can be derived for small $\delta frac_p$ (about 10%) with gradient bounds.[273] An error of $\delta \underline{a}$ causes a maximal change of the predicted value $|\mathcal{B}|$ of:

$$|\mathcal{B}| \leq \sum_{i=1}^{R} \left| \frac{\partial f(a)}{\partial a_i} \right| \cdot \delta a_i$$

If a flexible manufacturing system is considered in which each part type can be produced on alternative routes through the system, the fraction $frac_p$ for each part type must be split further to obtain the routing fractions q_{pr} each part type p has on route r. It is assumed, that a change $\delta frac_p$ will be split equally among the R_p alternative routes of part type p, i.e.:[274]

$$\delta q_{rp} = \frac{\delta frac_p}{R_p}$$

An approximation for the throughput change ΔT_{sys} of the system from a change of fractions $\delta frac_p$ can now be described by:

$$|\Delta T_{sys}| \leq \sum_{p=1}^{P} \sum_{r=1}^{R} \left| \frac{\partial T(q)}{\partial q_{rp}} \right| \cdot \frac{\delta frac_p}{R_p}$$

The calculation of the gradient for the production rate $T(q)$ can be obtained by closed queueing network theory (see Appendix B).

Appendix D:

Test runs for model EQUP-NEW-(1)

Several test runs were made with different problem sizes and changing part data on an IBM AT with 8 Mhz. The possible number of machine types M from which to chose was varied from 4 to 12 in pairs. For each size of M the number of part types was given by 4, 8 and 12 and the number of alternative routes for each part type by 2 and 5. Thus 24

---

273   see Tay, Y.C., Suri, R.: Error Bounds for Performance Prediction in Queueing Networks, in: ACM Trans. Comp. Syst., 3(1985)3, pp.227-254; Appendix B of that paper gives a mathematical interpretation of gradient bounds, which are based on first order approximation.

274   This assumption is only correct, if the routing fractions q are optimal with regard to the throughput. Then the gradients of the throughput on each route of a part type are equal (see results for the reciprocally related delay function in chapter 8.1.2.3), and it makes no difference which routing fraction of a part type p is changed.

different problems were tested and different solution procedures compared, of which the results are given in tab.D.1 and tab.D.2. Note that the solution time for the b&b algorithm using Lagrangian relaxation is in general longer than for the standard b&b algorithm (see tab.D.2). However, the former allows any increasing concave cost function, whereas the latter is restricted to linear cost functions.

number of	Branch and Bound (exact solut.)			Greedy heuristic		differences	
machines/part -types/routes	numb. iter.	solut. time [h:min:sec]	syst. costs F_{ex}	sol. time [sec]	sy. costs F_{heu}	absolute $F_{heu}-F_{ex}$	relat.
4/4/2	485	0:02:45	2370390	0,16	2406772	36382	1,5%
4/4/5	211	0:02:34	2029410	0,33	2725678	696268	34,3%
4/8/2	198	0:02:41	4281070	0,55	5505659	1224589	28,6%
4/8/5	503	0:12:35	3541650	1,15	5173609	1631959	46,0%
4/12/2	304	0:06:20	5075200	1,04	6261680	1186480	23,4%
4/12/5	144	0:05:36	4794950	2,63	6809725	2014775	42,0%
aver. values	308	0:05:25	3682112	0,98	4813854	1131742	30,7%
6/4/2	3382	0:34:06	5397430	0,33	6380331	982901	18,2%
6/4/5	1216	0:22:29	4152270	0,61	5576661	1424391	34,3%
6/8/2	789	0:14:12	8680830	0,99	9965988	1285158	14,8%
6/8/5	318	0:09:50	8269680	2,31	9639776	1370096	16,6%
6/12/2	882	0:19:58	13481350	2,58	16652880	3171530	23,5%
6/12/5	235	0:10:27	12428110	6,31	15890940	3462830	27,9%
aver. values	1137	0:18:30	8734945	2,19	10684429	1949484	22,3%
8/4/2	780	0: 9:47	7298520	0,44	7670337	371817	5,1%
8/4/5	2215	0:48:27	6228760	1,10	6947326	718566	11,5%
8/8/2	1482	0:26:27	14495610	1,87	15173320	677710	4,7%
8/8/5	611	0:22:30	11763810	3,90	13588360	1824550	15,6%
8/12/2	1025	0:24:56	18239820	3,63	19796470	1556650	8,5%
8/12/5	3076	2:49:53	17595800	7,58	21158800	3563000	20,2%
aver. values	1531	0:50:20	12603720	3,09	14055768	452048	11,6%
10/4/2	2519	0:54:23	9858700	0,66	10763690	904990	9,2%
10/4/5	16141	10:55:09	8200550	1,20	10134490	1933940	23,6%
10/8/2	6090	4:01:18	17308710	2,42	20039230	2730520	15,8%
10/8/5	11070	7:39:50	13805770	4,56	16782290	2976520	21,6%
10/12/2	4346	2:04:08	21982960	3,68	24646770	2663810	12,1%
10/12/5	2924	3:24:27	19479200	6,86	24082210	4603010	23,6%
aver. values	7182	4:49:52	15105981	3,23	17741446	2635465	17,4%
12/4/2	59992	24:00:17	11667300	0,77	12708830	1041530	8,9%
12/4/5	27094	20:05:03	9744560	1,59	11759520	2014960	20,7%
12/8/2	14064	5:16:59	19708760	2,63	21275980	1567220	7,6%
12/8/5	8747	11:41:48	16827650	5,11	20063320	3235670	19,2%
12/12/2	6181	6:33:42	26510690	5,77	28862910	2352220	8,8%
12/12/5	1935	2:36:03	23616680	9,12	27630660	4013980	17,0%
aver. values	19669	11:42:18	18012606	4,16	20383536	2370930	13,1%
	5963	3:33:17	11627873	2,73	13535807	1907934	19,0%

tab. D.1: Test results of model EQUP-NEW-(1)

number of	time for different b&b algorithms [h:min:sec]	
machines/part -types/routes	standard b&b alg.[275]	b&b using Lagrangian relaxation
4/4/2	0:02:45	00:00:05
4/4/5	0:02:34	00:01:09
4/8/2	0:02:41	00:03:15
4/8/5	0:12:35	00:36:50
4/12/2	0:06:20	00:05:23
4/12/5	0:05:36	05:49:58
aver. values	0:05:25	01:06:07
6/4/2	0:34:06	00:00:41
6/4/5	0:22:29	00:23:25
6/8/2	0:14:12	00:07:44
6/8/5	0:09:50	10:20:25
6/12/2	0:19:58	01:25:49
6/12/5	0:10:27	17:17:00
aver. values	0:18:30	04:55:55
8/4/2	0: 9:47	00:01:57
8/4/5	0:48:27	03:43:48
8/8/2	0:26:27	05:45:50
8/8/5	0:22:30	16:47:38
8/12/2	0:24:56	30:06:31
8/12/5	2:49:53	16:46:33
aver. values	0:50:20	12:12:13
10/4/2	0:54:23	00:01:23
10/4/5	10:55:09	
10/8/2	4:01:18	27:42:48
10/8/5	7:39:50	16:43:03
10/12/2	2:04:08	16:39:52
10/12/5	3:24:27	16:49:20
aver. values	4:49:52	15:35:44
12/4/2	24:00:17	01:26:54
12/4/5	20:05:03	84:20:08
12/8/2	5:16:59	44:04:29
12/8/5	11:41:48	43:39:10
12/12/2	6:33:42	43:49:15
12/12/5	2:36:03	65:59:23
aver. values	11:42:18	47:13:13
	3:33:17	16:12:37

tab. D.2: Test results of model EQUP-NEW-(1)

---

275   b&b code taken from Kuester, J.L., Mize, J.H.: Optimization Techniques with Fortran, New York 1973

Appendix E:

FIGURES

TABLES

ALGORITHMS

Appendix F:

ABBREVIATIONS

alg.	:	algorithm
CIRP	:	Centre Intersyndical d'Etudes et de Recherche sur la Productivité
Diss.	:	dissertation
e.g.	:	for example
Ed.	:	editor
fig.	:	figure
i.e.	:	that is
IEEE	:	Institute of Electrical and Electronics Engineers
Int.	:	International
Proc.	:	Proceedings
tab.	:	table

JOURNALS

ACM Perf. Eval. Rev.	:	Association for Computing Machinery Performance Evaluation Review
ACM Trans. Comp. Syst.	:	Association for Computing Machinery Transactions on Computer Systems
AIIE Trans.	:	American Institute of Industrial Engineers Transactions
Annals of OR	:	Annals of Operations Research
APII	:	Automatique - Productique Informatique Industrielle
Comm. ACM	:	Communications of the Association for Computing Machinery
Computers ind. Engng.	:	Computers and industrial Engineering
Comp. & Ops. Res.	:	Computers & Operations Research
DBW	:	Der Betriebswirt
EJOR	:	European Journal of Operations Research
HBR	:	Harvard Business Review
IEEE Trans. on Computers	:	Institute of Electrical and Electronics Engineers Transaction on Computers
IIE Trans.	:	Institute of Industrial Engineers Transactions
IJOPM	:	International Journal of Operations and Production Management
IJPR	:	International Journal of Production Research
INFOR	:	INFOR Journal, Canadian Journal of Operations Research and Information Processing
J. ACM	:	Journal of the Association for Computing Machinery
J. Appl. Prob.	:	Journal of Applied Probability
J. Manuf. Syst.	:	Journal of Manufacturing Systems
J. Opl. Res. Soc.	:	Journal of the Operational Research Society
JOTA	:	Journal of Optimization Theory and Applications
MP	:	Mathematical Programming
OMEGA	:	OMEGA The International Journal of Management Science
OR	:	Operations Research
Proc. Instn. Mech. Engrs.	:	Proceedings of the Institution of Mechanical Engineers
VDI-Z	:	Verein Deutscher Ingenieure - Zeitung
ZfB	:	Zeitschrift für Betriebswirtschaft

DEFINITIONS OF INDICES, PARAMETERS AND VARIABLES

Indices:

b	:	index for number of servers (machines, set-up tables, carriers etc.) at cell
c	:	index for the pallet type
d	:	system index
g	:	part group index
h	:	transportation system index
i	:	part family index
j	:	index for task j
k	:	index for operation or task k
m	:	index for cells, transportation system or assembly station
p	:	part type index
r	:	index for routes
t	:	time index
v	:	index to distinguish individual machines within a class

Parameters and Variables:

$\#[A]$	:	number of elements of the set A
α	:	Lagrange multiplier
$B(s_m, \rho_m)$	:	blocking probability
bl_m	:	limit for blocking probability at cell m
B_p	:	annualized costs savings if part p is produced on a flexible manufacturing system
BSU_{id}	:	annual batch set up costs if part family i is assigned to system d
BU	:	limited budget available
CC_p	:	capital costs for part type p
CF_{grt}	:	discounted payments for operating for machine group g on route r in period t
CL_{mpr}	:	labor costs of part type p at machine m if route r is used
CLU_{mk}	:	costs for load/unload operations at machine m for task k
C_m	:	operating and capital investment costs per server
C_N	:	operating and capital investment costs per pallet
COC_h	:	annual computer costs when material handling system m is used
CO_{gr}	:	operating costs for part group g on route r
$CO_{m,jk}$	:	operating costs of task j at station m when task k with $k > j$ is also performed on station m
CO_{mk}	:	operating costs for task k at machine m
CO_{mpr}	:	operating costs (or payments) of part type p on route r at machine m
CT_{mpr}	:	tool costs of part type p at machine m if route r is used
d_c	:	fraction that the throughput of pallet type c has on the system's throughput
$D_c(N)$	:	throughput time for pallet type c
δ_m	:	binary variable which equals one if station m is chosen to be included in the system, zero if not
D_{max}	:	maximal throughput time
D_N	:	overall average delay of a pallet in a system with N pallets
ϵ	:	return-on-investment correction factor
e_b	:	relative flow to the bottleneck-station (bottleneck-cell) m
EF_m	:	efficiency factor for machines/transport vehicles of type m
EF_{md}	:	efficiency factor on machine type m when assigned to system d
e_m	:	relative flow to station (cell) m
F_{bmt}	:	discounted payments for equipment (cell or transportation system) of type m in period t consisting of b servers (machines, load/unload stations, vehicles etc.)
FFE_m	:	payments for equipment expansion of type m cell or transportation system

FF_m	:	payments for equipment of a type m cell or transportation system in the first period
$f_m(T_m(n_m))$	:	profit function for cell m with throughput T_m and n_m buffers
$f_m(T_m)$	:	profit at cell m as a function of its throughput T_m
$frac_g$	:	product rate, i.e. the fráction part group g has of total production
$frac_{gt}$	:	product rate, i.e. the fraction part group g has of total production in period t
$frac_p$	:	product rate, i.e. the fraction part p has of total production
FVE_m	:	payments for equipment expansion of type m server (machines, load/unload stations, vehicles etc.)
FV_m	:	payments for equipment of a type m server (machines, load/unload stations, vehicles etc.) in the first period
$g_m(n_m)$	:	cost function for the allocation of n_m buffers
$g_m(s_m)$	:	cost at cell m as a function of the number of servers s_m allocated to it
i	:	rate of return
I_{bm}	:	equipment costs for cell or transportation system having b machines or b vehicles respectively
IFI_p	:	fixture costs per period for part type p
IF_m	:	fixed cell costs or fixed costs for the transportation system per period
I_m	:	capital costs for a station (cell) of type m
IP	:	costs per period for one pallet
ISU_{id}	:	initial set-up costs for part familiiy k at system d
IV_m	:	variable cell costs or variable costs for the transportation system for each additional machine, load/unload station, vehicle etc.
IY_{id}	:	annual inventory costs
K	:	production time per period
K_m	:	available capacity of machine type m or of arc m
$k_m(s_m)$	:	convex cost function for a cell of type m, dependent on the number of servers s_m
LF	:	Lagrangian Function
LF_a	:	Augmented Lagrangian Function
$\lambda_m(k)$	:	arrival rate to cell m, given the number of jobs there is k, $0 < \lambda_m(k) < \infty$ if $k < N_m$ and $\lambda_m(k) = 0$ if $k \geq N_m$
I_{mk}	:	annual load/unload time of task k at machine m
LN	:	large number
I_{pr}	:	route length of part type p on route r
M	:	amount of cells or amount of cells plus transportation system
$MHSC_h$	:	annual system costs when material handling system h is used
MHS_{ms}	:	binary variable assigning the material handling system of type h to system d
$MHWC_h$	:	annual labor costs per unit time of the handler
μ_m	:	service rate at cell m
MOC_{hd}	:	annual labor costs for machine operators for material handling system h of system d
MOC_{md}	:	annual labor costs for machine operators for machine type m of system d
N	:	(maximal) number of parts in system (inprocess inventory)
N^b	:	lower bound for the number of pallets
N_c	:	number of pallets of type c in the system, $c = \{1,..,C\}$
N_m	:	buffer limit at cell m
n_m	:	equilibrium number of parts at cell m, i.e. queue length at station m
N_{max}	:	maximal number of pallets
NS	:	number of systems
$\tilde{n}_m$	:	average queue length at cell m
N_+	:	ABA-bound estimate of N
OH_{hd}	:	annual overhead costs when material handling system h is used for system d
PMIN	:	expected lifetime of the system
qc_{ij}	:	required quantity of type (i,j) commodity at node j from node i
q_{pr}	:	fraction of part flow on route r for part p
R	:	production rate of the upstream production stage
R_{min}	:	required production rate for the system
R_{minc}	:	minimal required production rate for pallet type c

r_{mp}	:	flow rate of type p pieces on arc m of the network
R_{pmax}	:	production demand of part type p
R_{tmin}	:	required production rate for the system in period t
S	:	amount of available (identical) machines
s_m	:	number of parallel servers at cell m or at the transportation system
s_{M+1}	:	number of pallets in the system
s_{md}	:	integer variable signifying the number of machines of type m assigned to system d
SMW_m	:	amount of workload so far assigned to cell/transportation system m
SR_g	:	production requirements for part group g
SS_m	:	tool slot capacity
st_{pk}	:	set of tools required to produce work segment k of part p
SW_0	:	system workload
SW_{min}	:	minimum workload of the system
T	:	throughput of the system
t'_{mp}	:	sum of annual production time of all part types p at machine tool of type m
$T(\underline{s})$	:	system throughput, dependent on the number of servers at each station
$T_c(\underline{N})$	:	throughput rate of the system for pallet type c
$T_c(\underline{s})$	:	throughput for pallet type c
t_{imd}	:	processing time required by part family i on machine type m when assigned to system d
t_m	:	average travel time on arc m, or average processing time at station (cell) m
T_m	:	throughput of cell m
$T_M(\underline{c}, N)$	:	throughput of the system with M stations (cells and transportation system)
$T_m(s_m)$	:	throughput of station m, given that s_m servers have been allocated to it
t_{mjk}	:	operating time of task j at station m when task k with k > j is also performed on station m
t_{mk}	:	annual operating time of task k at machine m
t_{mpk}	:	processing time for part p at machine m for operation k
t_{mpr}	:	processing time for part type p on route r at machine m
T_N	:	system throughput with N pallets
T_p	:	throughput of part p
t_{pkm}	:	production time for work segment k of part type p at machine type m
ts_{mc}	:	sojourn time of pallet type c at cell (station) m
U_m	:	utilization of machine m
v_{mpr}	:	number of visits for part type p on route r at machine m
w_1	:	weight factor for equality constraints
w_2	:	weight factor for inequality constraints
W_{gmr}	:	workload of part group g at cell m on route r
W_m	:	workload at cell m
W_{mpr}	:	workload of part type p on route r at cell m
x_{gr}	:	production rate of part group g produced on route r
$x_{id}(i,j)$	:	binary variable, which assigns part family i to system d
x_{kl}	:	flow from node k to node l between the source node i and sink node j
x_m	:	input rate at cell m
x_{mjk}	:	binary variable equals one if task k follows task j at station m.
x_{mk}	:	fraction of annual volume of task k assigned to station m.
x_{mpk}	:	throughput at machine m of operation k for part type p
X_p	:	binary variable which equals one if part p is produced on the flexible manufacturing system and equals zero if not.
x_{pkmv}	:	binary variable which assigns segment k of part p to an individual machine v of type m
x_{pr}	:	production flow for part p on route r
x_{rgt}	:	production rate of part p produced on route r in period t
y_m	:	binary variable denoting the inclusion of station m in the assembly system configuration.
Y_m	:	binary variable which is one, if cell (machine type) or the transportation system of type m is included in the flexible manufacturing system
y_{mk}	:	fraction of task k that must be loaded onto machine m.

z_{bm} :   binary integer variable which is one if cell (transportation system) m with the amount of b machines (vehicles) is included in flexible manufacturing system.

z_{bmt} :   binary integer variable which is one if cell (transportation system) m with the amount of b machines (vehicles) is included in the flexible manufacturing system in period t.

Z^+ :   the set of positive integers

Z^+_0 :   the set of nonnegative integers

References

Adams, F.P., Cox, J.F.: Manufacturing Resource Planning: An Information Systems Model, in: Long Range Planning, 18(1985)2, pp.86-92

Adler, P.S.: A Plant Productivity Measure for "High-Tech" Manufacturing, in: Interfaces, 17(1987)6, pp.75-82

Aneke, N.A.G., Carrie, A.S.: A design technique for the layout of multi-product flowlines, in: IJPR, 24(1986)3, pp.471-481

Arbeitskreis "Langfristige Unternehmensplanung" der Schmalenbach Gesellschaft: Strategische Planung, in: Planung und Kontrolle, editor: H. Steinmann, München 1981, pp.23-45

Askin, R.G., Subramanian, S.P.: A cost-based heuristic for group technology configuration, in: IJPR, 25(1987)1, pp.101-113

Assad, A.A., Golden, B.L.: Expert Systems, Microcomputers, and Operations Research, in: Comp. & Ops. Res., 13(1986)2/3, pp.301-321

Avonts, L.H., Gelders, L.F., Van Wassenhove, L.N.: Allocation work between an FMS and a conventional jobshop: A case study, in: EJOR, 33(1988), pp.245-256

Ballakur, A.: An Investigation of Part Family/Machine Group Formation for Designing Cellular Manufacturing Systems, Ph.D. Thesis University Wisconsin, Madison 1985

Banks, J., Carson, J.S.: Discrete-Event System Simulation, Englewood Cliffs 1984

Baskett, F., Chandy, K.M., Muntz, R.R., Palacios, F.G.: Open, Closed and Mixed Netwoks of Queues with Different Classes of Customers, in: J.ACM, 22(1975)2, pp.248-260

Bazaraa, M.S., Shetty, C.M.: Nonlinear Programming: Theory and Algorithms, New York 1979

Bondi, A.B., Whitt, W.: The Influence of Service-Time Variability in a Closed Queueing Network of Queues, in: Performance Evaluation, 6(1986), pp.219-234

Brassard, G., Bratley, P.: Algorithmics Theory and Practice, Englewood Cliffs 1988

Browne, J., Dubois, D., Rathmill, K., Sethi, S.P., Stecke, K.E.: Classification of flexible manufacturing systems, in: The FMS Magazine, April 1984, pp.114-117

Bruell, S.C., Balbo, G.: Computational Algorithms for Closed Queueing Networks, New York and Oxford 1980

Burbidge, J.L.: Production Flow Analysis, in: The Production Engineer, 42(1963)12, pp.742-752

Burstein, M.C., Talbi, M.: Economic Evaluation for the optimal Introduction of Flexible Manufacturing Technology under Rivalry, in: Annals of OR, 3(1985), pp.81-112

Buzacott, J. A.: The fundamental principles of flexibility in manufacturing systems, in: Proc. of the 1st Intern. Conf. on Flexible Manufacturing Systems, Brighton U.K. 1982, pp.13-22

Buzacott, J.A., Shanthikumar, J.G.: Models for Understanding Flexible Manufacturing Systems, in: AIIE Trans., 12(1980)4, pp.339-350

Buzen, J.: Computational Algorithms for Closed Queueing Networks with Exponential Servers, in: Comm. ACM, 16(1973)9, pp.527-531

Cantor, D.G., Gerla, M.: Optimal Routing in a Packet Switched Computer Network, in: IEEE Trans. on Computers, C-23(1974)10, pp.1062-1069

Carrie, A.: Simulation of Manufacturing Systems, Chichester 1988

Carrie, A.S.: Numerical Taxonomy Applied to Group Technology and Plant Layout, in: IJPR, 11(1973)4, pp.399-416

Carrie, A.S.: The layout of multi-product lines, in: IJPR, 13(1975)6, pp.541-557

Carter, M.F.: Designing flexibility into automated manufacturing systems, in: Proc. 2nd ORSA/TIMS Conf. on Flexible Manufacturing Systems: Operations Research Models and Applications, Ed.: Stecke, K.E., Suri, R., Amsterdam 1986, pp.107-118

Chakravarty, A.K., Shtub, A.: An integrated layout for group technology with in-process inventory costs, in: IJPR, 22(1984)3, pp.431-442

Chew, B.W.: No-Nonsens Guide to Measuring Productivity, in: HBR, (1988)1, pp.110-118

Cohen, G., Moller, P., Quadrat, J.P., Viot, M.: Linear System Theory for Discrete Event Systems, in: Proc. 23nd IEEE Conf. on Decision and Control, Las Vegas 1984, pp.539-544

Dallery, Y., Dubois, D.: L'analyse opérationelle: une approche non stochastique des systèmes de files d'attente, in: APII, 20(1986)1, pp.43-86

Dallery, Y., Frein, Y.: An efficient Method to determine the optimal Configuration of a Flexible Manufacturing System, in: Proc. 2nd ORSA/TIMS Conf. on Flexible Manufacturing Systems: Operations Research Models and Applications, Ed.: K.E. Stecke and R. Suri, Amsterdam 1986, pp.269-282

Dallery, Y., Frein, Y.: An Efficient Method to determine the Optimal Configuration of a Flexible Manufacturing System, in: Annals or OR, 15(1988), pp.207-225

Dallery, Y.: On modelling flexible manufacturing systems using closed queueing networks, in: Large Scale Systems 11(1986), pp.106-119

Denning, P.J., Buzen, J.P.: The Operational Analysis of Queueing Network Models, in: Computing Survey, 10(1978)3, pp.225-261

Drees, S., Gomm, D., Plünnecke, H., Reisig, W., Walter, R.: Bibliography of Petri Nets, in: Lecture Notes on Computer Sciences, No.266, Heidelberg, 1987, pp.309-451

Dupont-Gatelmand, K.: A survey of flexible manufacturing systems, in: J. Manuf. Syst., 1(1981)1, pp.1-16

Eager, D.L., Sevcik, K.C.: Performance Bound Hierarchies for Queueing Networks, in: ACM Trans. Comp. Syst., 1(1983), pp.99-115

Eiselt, H.A., Frajer von, H.: Operations Research Handbook, Standard algorithms and Methods, Berlin, New York 1977

El-Essawy, I.F.K.: The Development of Component Flow Analysis as a Production Systems' Design for Multi-Product Engineering Companies, Ph.D. Thesis, UMIST, U.K. 1971

Endesfelder, T., Tempelmeier, H.: The SIMAN Module Processor - A Flexible Software Tool for the Generation of SIMAN Simulation Models, in: Simulation in CIM and Artificial Intelligence Techniques, Ed.: Retti, J., Wichmann, K.E., Proc. European Simulation Multiconference, July 1987, Vienna

Erkes, K., Schmidt, H.: Flexible Fertigung, in: VDI-Z, 126(1984)15/16, pp.577-591

Erlenkotter, D.: A comparative study of approaches to dynamic location problems, in: EJOR, 6(1981), pp.133-143

Falkner, C.H.: Flexibility in manufacturing plants, in: Proc. 2nd ORSA/TIMS Conf. on Flexible Manufacturing Systems: Operations Research Models and Applications, Ed.: Stecke, K.E., Suri, R., Amsterdam 1986, pp.95-106

Fiacco, A.V., McCormick, G.B.: Nonlinear Programming: Sequential Unconstrained Minimization Techniques, New York 1968

Fisher, M.L., Northup, W.D., Shapiro, J.F.: Using Duality to solve Discrete Optimization Problems: Theory and Computational Experience, in: Mathematical Programming Study, 3(1975), pp.56-94

Fisher, M.L.: The Lagrangian Relaxation Method for solving Integer Programming Problems, in: Management Science, 27(1981)1, pp.1-18

Fletcher, R.: Methods for Nonlinear Constraints, in: Nonlinear Optimization 1981, Ed.: M.J.D. Powell, London, New York 1982, pp.185-211

Fox, B.: Discrete Optimization via Marginal Analysis, in: MS, 13(1966), pp.210-216

Fratta, L., Gerla, M., Kleinrock, L.: The Flow Deviation Method: An Approach to Store-and-Forward Communication Network Design, in: Networks, 3(1973), pp.97-133

Geoffrion, A.M.: Lagrangian Relaxation for Integer Prgramming, in: Mathematical Programming Study, 2(1974), pp.82-114

Gerla, M.: The design of store-and-forward (S/F) networks for computer communications, Ph.D. thesis Dep. Comp. Science, Univ. Calif. Los Angeles, 1973

Gerwin, D.: An agenda for research on the flexibility of manufacturing processes, in: IJOPM, 7(1987)1, pp.38-49

Gordon, K.D., Dowdy, L.W.: The Impact of Certain Parameter Estimation Errors in Queueing Network Models, in: ACM Perf. Eval. Rev., 2(May 1980) pp.3-9

Gordon, W.J., Newell, G.F.: Closed Queueing Systems with Exponential Servers, in: OR, 15(1967)2,pp.254-265

Graves, S.C., Lamar, B.W.: A Mathematical Programming Procedure for Manufacturing System Design and Evaluation, in: Proc. IEEE Int. Conf. on Cir. and Comput., 1980, p.1146-1149

Graves, S.C., Lamar, B.W.: An Integer Programming Procedure for Assembly System Design Problems, in: OR, 31(1983)3, pp.522-545

Graves, S.C., Whitney, D.E.: A Mathematical Programming Procedure for Equipment Selection and System Evaluation in Programming Assembly, in: Proc. 18th IEEE Conf. on Decision and Control, Fort Lauderdale 1979, pp.531-534

Hahn, D.: Planung- und Kontrollrechnung, PuK, Wiesbaden 1974, p.64-67

Hayes-Roth, F., Waterman, D.A., Lenat, D.B.: Building Expert Systems, Reading Mass. 1983

Held, M., Karp, R.M.: The traveling salesman problem and minimum spanning trees, in: Mathematical Programming, 1(1971), pp.6-25

Held, M., Wolfe, P., Crowder, H.D.: Validation of subgradient optimization, in: Mathematical Programming, 6(1974), pp.62-88

Hestenes, M.R.: Multiplier and Gradient Methods, in: JOTA, 4(1969), pp.303-320

Hyer, N.L., Wemmerlöv, U.: Group Technology oriented Coding Systems: Structures, Applications, and Implementation, in: Production and Inventory Management, (1985)2, pp.55-78

Hildebrant, R.R.: Scheduling Flexible Machining Systems using Mean Value Analysis, in: Proc. IEEE Conf. on Decision and Control, Albuquerque, New Mexico 1980, pp.701-706

Ho, Y.C.: Perturbation Analysis explained, in: Working Paper Devision of Applied Sciences, Harvard University, 1987, p.1

Horváth, P., Kleiner, F., Mayer, R.: Dynamische Investitionsrechnung für flexibel automatisierte Werkzeugmaschinen, in: DBW, 47(1987)1, pp.69-84

Huang, P.Y., Houck, B.L.W.: Cellular Manufacturing: An Overview and Bibliography, in: Production and Inventory Management, (1985)4, pp.83-93

Hutchinson, G.K., Holland, J.R.: The Economic Value of Flexible Automation, in: J. Manuf. Syst., 1(1982)2, pp.215-218

Jackson, J.: Networks of Waiting Lines, in: OR, 5(1957),pp.518-521

Jackson, J.R.: Jobshop-Like Queueing Systems, in: MS, 10(1963), pp.131-142

Jacob, H.: Unsicherheit und Flexibilität: Zur Theorie der Planung bei Unsicherheit, in: ZfB, 44(1974)5, pp.299-326

Jacobsen, S.K.: Heuristic solution to dynamic plant location problems, in: M. Roubens (Ed.), Advances in Operation Research, Amsterdam 1977, pp.207-211

Kalkunte, M.V., Sarin, S.C., Wilhelm, W.E.: Flexible Manufacturing Systems: A Review of Modelling Approaches for Design, Justification and Operation, in: Flexible Manufacturing Systems: Methods and Studies, Ed.: A. Kusiak, Amsterdam 1986, pp.3-25

Kenevan, J.R., Mayrhauser von, A.K.: Convexity and Concavity Properties of Analytic Queueing Models for Computer Systems, in: Performance '84, Ed.: E. Gelenbe, Amsterdam 1984, pp.361-375

Kimemia, J.G., Gershwin, S.B.: Flow Optimization in Flexible Manufacturing Systems, in: IJPR 23(1985)1, pp.81-96

Kimemia, J.G., Gershwin, S.B.: Multicommodity Network Flow Optimization in Flexible Manufacturing Systems, in: Complex Materials Handling and Assembly Systems Final Report Vol. II No. ESL-FR-834-2, Electr. Syst. Lab. M.I.T., Cambridge MA, July 1978

Kimemia, J.G., Gershwin, S.B.: Network Flow Optimization in Flexible Manufacturing Systems, in: Proc. IEEE Conf. on Decision and Control 1979 pp.633-639

King, J.R., Nakornchai, V.: Machine-Component Group Formation in Group Technology: Review and Extension, in: IJPR, 20(1982)2, pp.117-133

King, J.R.: Machine-Component Group Formation in Group Technology, presented at the Vth Intern. Conf. on Production Research, Amsterdam Aug. 1979, in: OMEGA, 8(1980)2, pp.193-199

King, J.R.: Machine-Component Grouping in Production Flow Analysis: An Approach Using Rank Order Clustering Algorithm, in: IJPR, 18(1980)2, pp.213-232

Kiran, A.S., Tansel, B.C.: The System Setup in FMS: Concepts an Formulation, in: Proceedings of the Second ORSA/TIMS Conference on Flexible Manufacturing Systems, Ed.: K.E. Stecke, R. Suri, Amsterdam 1986, 321-332

Kleinrock, L. Queueing Systems, Vol.1 New York, Toronto 1976

Kleinrock, L. Queueing Systems, Vol.2 New York, Toronto 1976

Kobayashi, H., Gerla, M.: Optimal Routing in Closed Queueing Networks, in: ACM Trans. Comp. Syst., 1(1983)4, pp.294-310

Kriz, J.: Knowledge-based systems in industry: Introduction, in: Knowledge-based systems in industry, editor J. Kriz, Chichester 1987, pp.11-16

Kuester, J.L., Mize, J.H.: Optimization Techniques with Fortran, New York 1973

Kulatilaka, N.: Capital Budgeting and Optimal Timing of Investments in Flexible Manufacturing Systems, in: Annals of OR, 3(1985), pp.35-57

Kusiak, A., Heragu, S.S.: Expert Systems and Optimization in Automated Manufacturing Systems, Working Paper No. 07/87, University of Manitoba, Departement of Mechanical and Industrial Engineering, May 1987

Kusiak, A.: Flexible Manufacturing Systems: a structural approach, in: IJPR, 23(1985)6, pp.1057-1073

Lavenberg, S.S., Reiser, M.: Stationary State Probabilities at Arrival Instants for Closed Queueing Networks with Multiple Types of Customers, in: J. Appl. Prob. 17(1980), pp.1048-1061

Law, M.A., Kelton, W.D.: Simulation Modeling and Analysis, New York 1982

Lazowska, E.D., Zahorjan, J., Graham, G.S., Sevcik, K.C.: Quantitative System Performance Computer System Analysis Using Queueing Network Models, Englewood Cliffs 1984

Lehmann, M.R.: Wirtschaftlichkeit, Produktivität und Rentabilität (I), in: ZfB, 28(1958), pp.537-557,614-620

Little, J.D.C.: A Proof of the Queueing Formula $L = \lambda \cdot W$, in: OR 9(1961), pp.383-387

Maier, K.: Die Flexibilität betrieblicher Leistungsprozesse, Frankfurt am Main 1982;

Mandelbaum, M.: Flexibility in decision making: an exploration and unification, Ph.D. Thesis, Dept. of Industrial Engineering, University of Toronto, Ont., 1978

Martinez, J., Alla, H., Silva, M.: Petri Nets for the specification of FMSs, in: Modelling and Design of Flexible Manufacturing Systems, edited by A. Kusiak, Amsterdam 1986, pp.389-406;

McAuley, J.: Machine Grouping for Efficient Production, in: The Production Engineer, 51(1972)2,pp.53-57

McCormick, W.T., Schweitzer, P.J., White, T.E.: Problem Decomposition and Data Recognition by a Clustering Technique, in: OR, 20(1972),pp.993-1009

Meffert,H.: Zum Problem der betriebswirtschaftlichen Flexibilität, in: ZfB, 39(1969), pp.779-800

Mellichamp, J.M., Wahab, A.F.A.: An expert system for FMS design, in: Simulation,48(1987)5,pp.201-208

Meredith, J.R., Suresh, N.C.: Justification techniques for advanced manufacturing technologies, in: IJPR, 24(1986)5, pp.1043-1057

Michael, G.J., Millen, R.A.: Economic Justification of Modern Computer-Based Factory Automation Equipment: A Status Report, in: Annals of OR, 3(1985), pp.25-34

Miltenburg, G.J., Krinsky, I.: Evaluating Flexible Manufacturing Systems, in: IIE Trans., 19(1987)2, pp.222-233

Molloy, M.K.: On the Integration of Delay and Throughput Measures in Distributed Processing Models, Ph.D. Thesis, University of California, Los Angeles 1981

Moore, C.G.: Network Models for Large-Scale Time-Sharing Systems, Technical Report No.71-1, Department of Industrial Engineering, University of Michigan, Ann Arbor 1971

Mosier, C., Taube, L.: The Facets of Group Technology and Their Impacts on Implementation - A State-of-the-Art Survey, in: OMEGA, 13(1985)5, pp.381-391

Mutti, R., Semeraro, Q.: A heuristic method to group pieces for economical production, Working Paper WP-1987-001-QS, Politecnico di Milano, dipartimento di meccanica, Milano 1987

Narahari, Y., Viswanadham, N.: A Petri Net Approach to the Modelling and Analysis of Flexible Manufacturing Systems, in: Annals of OR, 3(1985), pp.449-472

Nelson, C.A.: A scoring model for flexible manufacturing systems project selection, in: EJOR, 24(1986), pp.346-359

NN: Ihre verantwortlichen Partner für erfolgreiche Fertigungsstätten, prospectus from Dörries Scharmann GmbH, Düren, Mönchengladbach 1988

Olivia-Lopez, E., Purcheck, G.F.: Load Balancing for Group Technology Planning and Control, in: Int. Journal of Machine Tool Design and Research, 19(1979)4, pp.259-274

Opitz, H.: Verschlüsselungsrichtlinien und Definitionen zum werkstückbeschreibenden Klassifizierungssystem, Essen 1966

Petri, C.A.: Kommunikation mit Automaten, Schriften der Rheinischen Westfälischen Instituts für instrumentelle Mathematik an der Universität Bonn, Bonn 1962

Powell, M.J.D.: A method for nonlinear constraints in minimization problems, in: Optimization, Ed.: R. Fletcher, New York 1969, pp.283-298

Powell, M.J.D.: Algorithms for Nonlinear Constraints that use Lagrangian Functions, in: MP, 14(1978), pp.224-248

Price, T.G.: Balanced Computer Systems, Stanford Electronics Lab. Technical Report No.88, Stanford University, Stanford April 1974 pp.35-49

Primrose, P.L.: Evaluating the 'intangible' benefits of flexible manufacturing systems by use of discounted cash flow algorithms within a comprehensive computer program, in: Proc. Instn. Mech. Engrs. Vol. 199 No.B1 (1985), pp.23-28

Pritsker, A.A.: The GASP IV Simulation Language, New York 1974

Rajagopalan, R., Betra, J.L.: Design of Cellular Production Systems: A Graph-Theoretic Approach, in: IJPR, 13(1975)6, pp.567-579

Ravichandran, R: Decision Support in FMS Using Timed Petri Nets, in: J. Manuf. Syst., 5(1986)2, pp.89-101

Reichwald, R., Behrbohm, P.: Flexibilität als Eigenschaft produktionswirtschaftlicher Systeme, in: ZfB, 53(1983)9, pp.831-851

Reiser, M., Lavenberg, S.S.: Mean-Value Analysis of Closed Multichain Queueing Networks, in: J.ACM, 27(1980)2, pp.313-322

Rockafellar, R.T.: A Dual Approach to Solving Nonlinear Programming Problems by Unconstrained Optimization, in: MP, 5(1973), pp.354-373

Rockafellar, R.T.: The Multiplier Method of Hestenes and Powell Applied to Convex Programming, in: JOTA, 12(1973)6, pp.558-561

Sarin, S.C., Chen, C.S.: A Mathematical Model for Manufacturing System Selection, in: Flexible Manufacturing Systems: Methods and Studies, Ed.: A. Kusiak, Elsevier Science Publishers B.V. (North Holland) 1986, pp.99-112

Schaefer, F.-W.: System zur Planung und Nutzung der Flexibilität in der Fertigung, Diss. der TH-Aachen, Aachen 1980

Schmitt, T.G., Klastorin, T., Shtub, A.: Production classification system: concepts, models and strategies, in: IJPR, 23(1985)3, pp.563-578

Schneeweiß, C.: Construction and selection of quantitative planning models - a general procedure illustrated with models for stock control, in: EJOR, 6(1981), pp.372-379

Schneeweiß, C.: Elemente einer Theorie betriebswirtschaftlicher Modellbildung, in: ZfB, 54(1984)5, pp.480-504

Schweitzer, P.J., Seidmann, A., Shalev-Oren, S.: The Corrections Terms in Approximate Mean Value Analysis, in: Operation Research Letters, 4(1986)5, pp.197-200

Schweitzer, P.J.: Approximate Analysis of Multiclass Closed Networks of Queues, in: Proc. Intern. Conf. on Stochastic Control and Optimization, Amsterdam 1979, pp.25-29

Secco-Suardo, G.: Optimization of closed queueing networks, in: Complex Materials Handling and Assembly Systems Final Report Vol. III No. ESL-FR-834-3, Electr. Syst. Lab. M.I.T., Cambridge MA, July 1978

Secco-Suardo, G.: Workload Optimization in a FMS Modelled as a Closed Network of Queues, in: Annals of the CIRP, 28(1979)1, pp.381-383

Seifoddini, H.: Cost based machine-component grouping model: in Group Technology, Ph.D. Thesis Oklahoma State University, Stillwater 1984

Shalev-Oren, S., Seidmann, A., Schweitzer, P.J.: Analysis of Flexible Manufacturing Systems with Priority Scheduling: PMVA, in: Annals of OR, 3(1985), pp.115-139

Shanthikumar, J.G., Yao, D.D.: On server allocation in multiple center manufacturing systems, in: OR, 36(1988)2, pp.333-342

Shanthikumar, J.G., Yao, D.D.: Optimal Buffer Allocation in a Multicell System, in: The International Journal of Flexible Manufacturing Systems, 1(1989), pp.347-356

Shanthikumar, J.G., Yao, D.D.: Optimal server allocation in a system of multi-server stations, in: MS, 33(1987)9, pp.1173-1180

Shanthikumar, J.G., Yao, D.D.: Second-Order Properties of the Throughput of a Closed Queueing Network, in: Math. OR, 13(1988)3, pp.524-534

Shanthikumar, J.G., Yao, D.D.: Second-Order Stochastic Properties in Queueing Systems, in: Proc. of the IEEE, 77(1989)1, pp.162-170

Shanthikumar, J.G., Yao, D.D.: Stochastic Monotonicity of the Queue-Lengths in Closed Queueing Networks, in: Operations Research, 35(1987)4, pp.583-588

Shanthikumar, J.G.: On the superiority of balanced load in a flexible manufacturing system, technical report, Department of IE & OR, Syracuse University, New York, 1982

Shanthikumar, J.G.: Stochastic Majorization of Random Variables with Proportional Equilibrium Rates, in: Adv. Appl. Prob., 19(1987), 854-872

Shapiro, J.F.: A Survey of Lagrangian Techniques for Discrete Optimization, in: Annals of Discrete Mathematics, 5(1979), pp.113-138

Solberg, J.J.: A Mathematical Model of Computerized Manufacturing Systems, in: Proc. 4th Intern. Conf. on Production Research, Tokyo, Japan, Aug.1977, pp.1265-1275

Solot, P., Bastos, J.M.: Choosing a Queueing Model for FMS, Working Paper O.R.W.P. 87/05, Ecole Polytechnique Fédéral de Lausanne, Département de Mathématiques, Lausanne 1987

Solot, P., Bastos, J.M.: MULTIQ: A Queueing Model for FMSs with Several Pallet Types, in: J. Opl. Res. Soc., 39(1988)9, pp.811-821

Solot, P.: Optimizing a Flexible Manufacturing System with several Pallet Types, Working Paper O.R.W.P. 87/17, Ecole Polytechnique Fédéral de Lausanne, Département de Mathématiques, Lausanne Sept. 1987

Stecke, K.E., Solberg, J.J.: The Optimal Planning of Computerized Manufacturing Systems, School of Industrial Engineering, Purdue University, Report No.20, West Lafayette, Indiana 1981

Stecke, K.E.: Design, Planning, Scheduling, and Control Problems of Flexible Manufacturing Systems, in: Annals of OR, 3(1985)3, pp.3-12

Stecke, K.E.: On the Nonconcavity of Throughput in Certain Closed Queueing Networks, in: Performance Evaluation, 6(1986), pp.293-305

Steudel, H.J., Ballakur, A.: A Dynamic Programming Based Heuristic for Machine Grouping in Manufacturing Cell Formation, in: Computers ind. Engng., 12(1987)3, pp.215-222

Suresh, N.C., Meredith, J.R.: Justifying Multimachine Systems: An Integrated Strategic

Approach, in: J. Manuf. Syst., 4(1985)2, pp.117-134

Suri, R., Cao, X.: Optimization of flexible manufacturing systems using new techniques in discrete event systems, in: Proc. 20th Allerton Conf. Communic. Control and Computing, Monticello, Illinois 1982, pp.434-443

Suri, R., Dille, J.W.: A Technique for on-line Sensitivity Analysis of Flexible Manufacturing Systems, in: Annals of OR, 3(1985), pp.381-391

Suri, R., Hildebrant, R.R.: Modelling Flexible Manufacturing Systems Using Mean Value Analysis, in: J. Manuf. Syst., 3(1984)1,pp.27-38

Suri, R., Whitney, C.: Decision Support Requirements in Flexible Manufacturing, in: J. Manuf. Syst., 3(1984)1, pp.61-69

Suri, R.: A Concept of Monotonicity and Its Characterization for Closed Queueing Networks, in: OR, 33(1985)3, pp.606-624

Suri, R.: An Overview of Evaluative Models for Flexible Manufacturing Systems, in: Annals of OR, 3(1985), pp.13-21

Suri, R.: Implementation of Sensitivity Calculations on a Monte Carlo Experiment, in: JOTA, 40(1983)4, pp.625-630

Tay, Y.C., Suri, R.: Error Bounds for Performance Prediction in Queueing Networks, in: ACM Trans. Comp. Syst., 3(1985)3, pp.227-254

Tempelmeier, H., Endesfelder, T.: Der SIMAN MODUL PROZESSOR - ein flexibles Softwaretool zur Erzeugung von SIMAN-Simulationsmodellen, in: Angewandte Informatik, 29(1987)2, pp.104-110

Tempelmeier, H., Kuhn, H., Tetzlaff, U.: Performance Evaluation of Flexible Manufacturing Systems with Blocking, in: IJPR, 27(1989)11, pp.1963-1979

Tempelmeier, H.: Kapazitätsplanung für flexible Fertigungssysteme, in: ZfB, 58(1988)9, pp.963-980

Van der Wal, J.: Monotonicity of the Throughput of a Closed Exponential Queueing Network in the Number of Jobs, in: OR Spektrum, 11(1989), pp.97-100

Van Looveren, A.J., Gelders, L.F., Van Wassenhove, L.N.: A Review of FMS Planning Models, in: Modelling and Design of Flexible Manufacturing Systems, Ed.: A. Kusiak, Amsterdam 1986, pp.3-31

Vinod, B., Solberg, J.J.: The optimal design of flexible manufacturing systems, in: IJPR, 23(1985)6, pp.1141-1151

Warnecke, H. J., Steinhilper, R.: Flexible manufacturing systems; new concepts; EDP-supported planning; application examples, in: Proc. 1st Intern. Conf. on Manufacturing Systems, Brighton U.K. 1982, pp.345-356

Weber, K., Trzebiner, R., Tempelmeier, H.: Simulation with GPSS - Lehr- und Handbuch with wirtschaftswissenschaftlichen Anwendungsbeispielen, Bern, Stuttgart 1983, pp.31-32

Whitney, C.K., Gaul , T.S.: Sequential Decision Procedures for Batching and Balancing in Flexible Manufacturing Systems, in: Annals of OR, 3(1985), pp.301-316

Whitney, C.K., Suri, R.: Algorithms for Part and Machine Selection in Flexible

Manufacturing Systems, in: Annals of OR, 3(1985), pp.239-261

Whitt, W.: Open and Closed Models for Networks of Queues, in: Bell Lab. Tech. J., 63(1984)9, pp.1911-1979

Wildemann, H.: Investitionsplanung und Wirtschaftlichkeitsrechnung für flexible Fertigungssysteme (FFS), Stuttgart 1987

Williams, A.C., Bhandiwad, R.A.: A Generation Function Approach to Queueing Network Analysis of Multiprogrammed Computers, in: Networks, 6(1976), pp.1-22

Yao, D.D., Kim, S.C.: Some Order Relations in Closed Networks of Queues with Multiserver Stations, in: Naval Research Logistics, 34(1987), pp.53-66

Yao, D.D., Shanthikumar, J.G.: Some resource allocation problems in multi-cell systems, in: Proc. 2nd ORSA/TIMS Conf. on Flexible Manufacturing Systems: Operations Research Models and Applications, Ed.: K.E. Stecke and R. Suri, Amsterdam 1986, pp.245-255

Yao, D.D., Shanthikumar, J.G.: The optimal input rates to a system of manufacturing cells, in: INFOR, 25(1987)1, pp.57-65

Zahorjan, J., Sevcik, K.C., Eager, D.L., Galler, B.: Balanced Job Bound Analysis of Queueing Networks, in: Comm. ACM, 25(1982)2, pp.134-141

Zäpfel, G.: Produktionswirtschaft, Berlin, New York 1982, pp.15-20

Zelenovic, D.M.: Flexibility - a condition for effective production systems, in: IJPR, 20(1982)3, S.319-337

Zimmermann,G.: Ergie keitsmaße der Produktion, in: Handwörterbuch der Produktionswirtschaft, Ed.: W. Kern, Stuttgart 1979, pp.520-528

Zoutendijk, G.: Mathematical Programming Methods, Amsterdam, New York, Oxford 1976